I0491682

We Are More Than Our Brains

Putting our minds to neuroscience

Trevor Rollings

We Are More Than Our Brains

Only connect

E M Forster

Printed by Kindle Direct Publishing

We Are More Than Our Brains

Also available on Amazon by the same author
in the series
Empires of the Mind

www.trevorrollings.com

The stupendous Story of Us
From Big Bang to Big Brother in fifteen giant leaps

No Man is an Island
How pain, suffering and hope bind us to each other

The Mosaic of the Mind
*Where does our mind go in sleep, dreams, hypnosis and
meditation?*

In Our Right Mind
Sex, drugs, kicks and control

Mountains of the Mind
How we get depth from flatness

The Mirror of the Mind
Seeing personally, seeing really and seeing sanely

The Smart Mind
A natural history of intelligence

Body, Mind and Machine
Staying human in the Age of AI

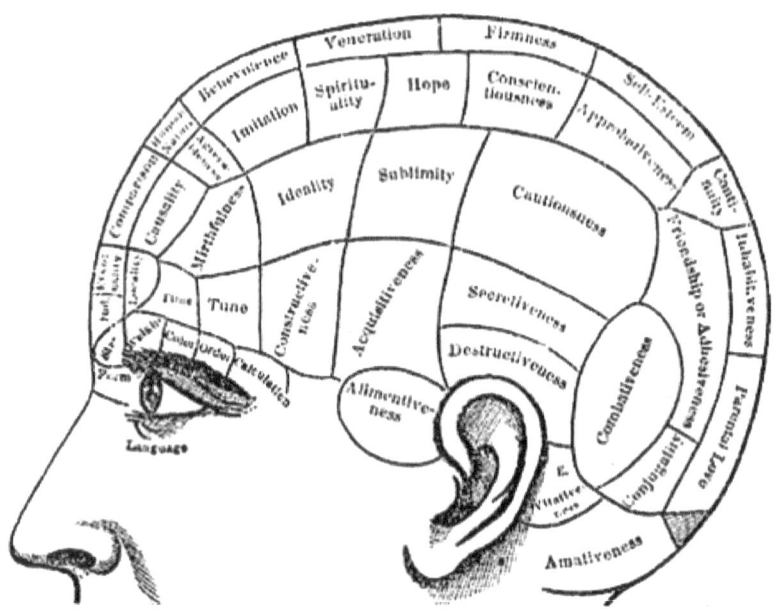

Looking in from the outside

The Ned Kelly Strategy: Phrenology
We are the bumps on our skull.
The Pavlov's Dog Strategy: Behaviourism
We are our surface actions.
The Fred Flintstone Strategy: Evolutionary Psychology
We are our ancestral genes.
The Alan Turing Strategy: Machine Learning
We are our programming and algorithms.
The Oedipus Strategy: Structuralism
We are the myths and memes we live by.

We Are More Than Our Brains

Peering out from the inside

The René Descartes Strategy: Rationalism
We are our free thoughts.
The Numskull Strategy: Cognitivism
We are flunkeys controlled by a chief executive.
The Phineas Gage Strategy: Neuroscience
We are our neurochemistry.
The Telecoms Strategy: Information Theory
We are our networked data.
The Therapeutic Strategy: Psychoanalysis
We are our memories and desires.

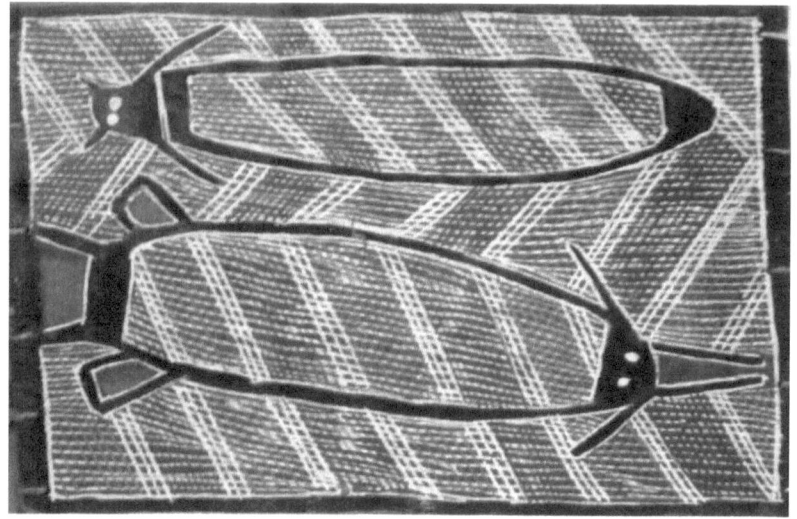

X-raying the brain

This unsigned Aborigine bark painting, purchased in Queensland in 1986, depicts two barramundi fish.

Its typical 'X-ray' style suggests that we can see into the inner workings of a living object, helping us to realise that, though fish are lower down the evolutionary scale than us, we share an ancestral brain design with them. Also, though they are swimming in opposite directions, like our brain and mind, they inhabit the same element.

This book explores these themes. To what extent can we peer inside our brain to see its operations, what light do our primitive origins throw on the evolution of our mind, and how are our brain and mind linked as complementary stories in a uniquely human contra-flow?

Table of Contents

We Are More Than Our Brains

We're Going on a Mind Hunt

We can't go over it.
We can't go under it.
Oh no!
We've got to go through it!

Michael Rosen

The neuroscientific revolution – philosophy of mind - reductionism – ways of knowing – choosing a narrative – Mr Popper's Three Worlds – the human life-world

- Neuroscience is an effective tool for unlocking how the brain works.
- As a biological organ, the brain can be investigated as the foundation of the mind.
- Our mind, which is our lived experience, cannot be investigated by this method.
- This is because we inhabit two worlds, each with its own way of seeing and knowing: physics and metaphysics, biology and being, things and thoughts.
- Between our ears is a unified mind-brain, a coalition of two distinct but complementary 'ways of being', neither solely explainable in terms of the other.
- Dualism, or splitting the brain and mind into two entities, is a neat way of resolving the mind-brain puzzle, but it creates more problems than it solves.
- Reducing mental events to material brain states doesn't work either, because it leaves half of our story untold. We are more than our brains.
- Using our brain to study itself is problematic, unless we constantly remind ourselves that, in our evolution, our mind-brain has been as strongly shaped by culture as by biology.

We Are More Than Our Brains

This is the Age of the Brain

We are privileged to live in the Age of the Brain. The splicing of the neuron has been as important a breakthrough in human knowledge as the splitting of the atom. Pioneering experiments and advanced technology have allowed scientists to conquer the final frontier of our ignorance: what goes on right behind our eyes.

Inspired by this new knowledge, neuroscientists and cognitive psychologists have revealed to us the mechanics of how our brain creates its picture of reality, the neuroscientist Dick Swaab boldly proclaiming 'We are our Brains' in his 2015 book of that title.

This is a catchy meme, but 'brainism' falls some way short of explaining what sort of beings we are. *We* are responsible for what we do, not our brain. Cut to the criminal standing in the dock, about to be sentenced. He turns to the judge and pleads, 'Go easy on me your honour, my brain made me do it'. He could just as easily blame his genes, hormones, karma, social media feed, or what he had for breakfast that morning.

Reducing all mental phenomena to electro-chemistry, gene expression and information tells only half of our story, that of 'becoming', the laws of which have been clearly laid out by science over the last four centuries. But there is another half to the story, that of 'being', with a parallel plot of who we think we are and how we relate to each other.

We can't tell one story without the other, nor is one more important. They are yin and yang, and in this book, we set out to discover why. We live within two narratives, or frames of reference, and we come a cropper when we muddle them up.

Even within the exciting new discipline of neuroscience, researchers realise that neurons do not work singly, but in complex networks that connect us to a wider society of minds. The human mind-brain (we'll explain the need for this unique pairing in due course) did not evolve singly, but in a community of organisms. Just as a bird builds a nest to protect its young from the harsh pressures of natural selection, so our species has grown culture to pass on to our young the life of the mind.

Biologists have found increasing evidence that organisms are not slavish carriers of their selfish genes, but play an active and cooperative role in their own evolution. Information theorists are

10

aware that, though our computers can be programmed to outstrip us at algorithmic logic, ontological realities such as emotion, consciousness, intentionality and agency cannot yet be coded into a robot.

The puzzle we are trying to solve

Biological and cognitive neuroscientists answer a twin call, to know how things work and to increase self-understanding. Their curiosity is as old as philosophy itself. Plato, and many thinkers after him, realised he could not answer questions about how the world works until he first understood how the mind operates. We can put nature on the rack, but who is doing the torturing, and why? We need to know the mind of the enquirer if we are to make any sense of the results of the enquiry.

Only then can we answer the 'how' questions. How do the firings of neurons, or brain events, result in the rich experience of mind and consciousness? What light do neuroscience's methods of enquiry throw on a much older humanist vision of ourselves, not as brains primed to keep us out of trouble and cut the best survival deal, but as minds looking for reasons and searching for values?

Neither scientists looking in from the outside, nor philosophers peering out from the inside, are able on their own to give us a full picture of ourselves. Our psychology is uneasily sandwiched between physiology and philosophy, each shining a different light on our self understanding.

Neuroscience shows how we can use our brain to study its own biology. We can use neuroimaging to trace organic causes of mental illnesses. It is a much greater challenge to use our brain to explain functional disorders of the mind. There may be nothing physically wrong with us, but we suffer psychologically nonetheless.

This is because we are the both the hunter and the hunted, unable to escape from the loop of our own minding. We are part of the puzzle we are trying to solve, so it is difficult to step outside the frame, or to study ourselves objectively. Some say that the mind will always remain a riddle wrapped in a mystery inside an enigma.

This is not the fault of brain science, which advances towards a linear and cumulative understanding, one small step at a time, predicated on repeated events and processes. The story of the mind is quite the opposite, replete with unpredictable setbacks and dilemmas, as culture, history and psychology show. Cosmic time rolls on without us, but human time is our awareness of the fragility of our lives, destined never to achieve complete self-understanding.

Animals can see their reflection in a mirror, but the gift of consciousness allows us to reflect on our own reflections, pondering our past and future. Introspection is a difficult art, offering no sure route to self knowledge. If the criminal could know the dark matter of his own mind, he might find himself walking a free man past the courthouse. But he suffers from a logical paradox: just as the eye cannot see itself, and the mind cannot be its own cause, so he cannot be his own judge and jury.

Watching our words
We expect too much of neuroscience if we think it can resolve all these conundrums, unveil ultimate reality or solve our political squabbles. Nevertheless, it has given us a new language and a new way of thinking about ourselves. To our familiar talk of brain waves, mental blocks, grey matter and senior moments, we can now add synaptic clean-ups, cognitive reboots, being too left-brained, and getting our neural wires crossed.

Though now part of the modern zeitgeist, the word 'brain' is a relative newcomer, not big enough by itself to carry the full burden of the human condition. It entered the English language about a thousand years ago, but emanates from a much older Greek root *phren*, from which we get words like phrenology, frenzy and frantic. It refers to the lump of flesh that sits between our ears, described in the dictionary as a convoluted mass of nervous substance. To this day, animal brains are eaten as a rich source of fat and protein.

It took many centuries of subtle and dedicated scientific endeavour to appreciate the brain's function as a biological organ, not to mention its role as an orchestrator of the intellect. To the biologist and evolutionist, the brain has evolved as a movement coordinator, food finder, memory storer, model maker, pattern detector, causal generator, future predictor and symbol processor.

We Are More Than Our Brains

These are vital roles for a neurologically complex brain like ours, but there is another term, mind, much older in origin, belonging to a family of words which includes measurement, memory and the goddess Minerva, overseer of thought and understanding.

In texts dating from over two thousand and a half years ago we begin to hear a new voice and sense of awareness, as if one consciousness is talking to another. It also goes by the name of soul, spirit, anima, heart, psyche, nous or atman. The word 'brain' does not cover this, whereas 'mind' embraces our lived experience, both as individual persons and social co-operators.

Two equal truths

For reasons which we shall explore in this book, we are more than our brains. We are our minds. This does not mean that the claim 'we are our brains' is untrue, or wrong. It is as true and right within the context of neuroscience as 'we are our minds' is appropriate when we talk psychology.

Shakespeare had no difficulty with this double-seeing, moving freely between encompassing the 'heat oppress'd brain' and 'the mind's construction' in a single thought, showing the dual insight of the scientist and humanist in the same moment. He understood that these are complementary truths, just as the quantum physicist realises that, though reality can appear as either a particle or a wave, both are present all of the time

In 1973 the evolutionary biologist Theodore Dobzhansky remarked, 'Nothing in biology makes sense except in the light of evolution'. A neuroscientist might tweak this to say that nothing in the mind makes sense except in the light of the brain. This is both an empirical and a material truth. But it is not the whole truth. To complete the picture, we need to add that nothing in the mind makes sense except in the light of the sculpting of our personal and social experience.

Without a brain, our body would die and our mind would evaporate, but without a mind, we have no means of contemplating our situation, or reason to do so. The clue is in the pronoun 'we': a brain is little use without an owner, and a mind has evolved to grow only in the company of other minds. 'We' allows us to progress from the idea of the brain as a computer built of neurons, to a mind possessed of beliefs and desires fashioned in a nurturing culture.

Theodore Dobzhansky
1900-1975
'Nothing in biology makes sense except in the light of evolution', said the eminent geneticist Dobzhansky. By similar reasoning, nothing in the mind makes sense except in the light of the brain. Can we also say that nothing in the brain makes sense except in the light of the mind?

Having a brain is a necessary condition for owning a mind, but not sufficient. We cannot say therefore that a brain 'causes' a mind. Even a simple decision to raise our arm is so far along a chain of causes that we never quite know when or why we are going to do it.

For centuries philosophers have puzzled over how the brain and mind relate to each other, and despite bold claims by some researchers, science has not single-handedly brought us any closer to an answer. This is partly because our species has inherited a 'gap instinct', or a tendency to think in binary terms. In the case of the mind-brain, this is an unhelpful false opposite that blocks deeper understanding. We don't have to choose between brain *or* mind, just as we don't have to choose between science *or* religion. Our mind-brain has grown to be at home in both.

Binary thinking is a particularly counter-productive strategy for unravelling human consciousness, referred to by philosophers of mind as the 'hard problem'. The physicist Niels Bohr noted that the opposite of a profound truth may well be another profound truth. To put it another way, 'we are our brains' is true only if 'we are our minds' is also true. Our brain has evolved not to function in glorious biological isolation, but in a rich milieu of mind, or other people.

The first neuroscientists were understandably single-focus and reliant on reductionism, if only to help them to narrow down what they were looking for. Today there is a broader synthesis, captured in the hybrid term mind-brain. This embraces the two equal truths of biology and experience, synapse and self, neuron and narrative, genes and culture, matter and spirit. If we want to see further, understand ourselves better and prepare for the challenges that our species faces, we must shine both lights on ourselves.

Biochemistry shapes our brain activity at the physical and material level, but mind operates at a metaphysical and existential level, requiring a different kind of explanation. When we're mentally ill, do we need a session with a shrink, a packet of pills, electro-shock therapy or brain surgery? Psychotherapy and psychiatry have been slugging this one out for a century, coming to a truce in the longest word they could find: psychoneuropharmacology.

Fortunately we have not evolved to *feel* we are living in parallel realities, one of neural activity and one of mental experience. We might talk of ourselves as if our mind and body are in competition, as in the phrase 'The spirit is willing but the flesh is weak'. But we don't lapse into dualism, otherwise we could never exercise mind over matter, go on a diet or remain celibate.

We need an inclusive vision of ourselves if we are to understand why we dream, can't sleep, have headaches, lose our short term memory as we age, hallucinate, need other people, fall victim to neural disorders that are hard to diagnose and treat, crave stimulus, fall prey to addiction, get depressed, grow suspicious of each other or, despite our fanciest technology, remain trapped in the wheel of suffering and the cycle of birth and death.

To get anywhere near an understanding of the human condition, we need an integrated concept of an embodied mind, or a mindful brain. As Democritus remarked at the very start of the Western scientific adventure, the truth lies in the depths.

We Are More Than Our Brains

Different ways of knowing

We also possess different ways of knowing, from the outside and from the inside. We flit each moment between a cool scientific 'take' on the world and the warm buzz of personal experience, poring sedately over a science text book one minute and freaking out in front of a horror movie the next.

As a result, we 'know' our brain in a completely different manner from the way we 'know' our mind. We feel that, though we *have* a brain, we *are* our mind. This is not just chopping logic or playing with words: it is a fundamental division in our cognition and experience. Although brain and mind both 'happen' between our ears, they perform very different jobs for us. No wonder the dictionary gives us contrasting but complementary definitions of each.

These differences of function between the mind and brain have nothing to do with whether we are 'sensory' types who need the evidence of our senses, or 'intuitives' who judge situations by how we feel. It has everything to do with the fact that we never act in the world as less than whole bodies and implicated minds.

Brain	Mind
A convoluted mass of soft nervous tissue in the skull of vertebrates at the top of the spinal column.	*What makes us aware of our self worth, social significance and cultural belonging.*
The coordinating centre of perception, sensation, intelligence and nervous activity.	*Our faculty of consciousness, idea-creation, intention, agency, feeling and reasoning.*
The organ of logic, prediction, gap-filling, risk-calculation and problem-solving.	*Our seat of memory, through which we write a narrative with our self at the centre.*
Note the focus on becoming, biology, nerves and cognition.	Note the emphasis on being, experience, values and identity.

We Are More Than Our Brains

We expect the word 'know' to cover our grasp of nature as perceived from the outside, and our thoughts as lived on the inside. This is a valuable distinction, but one that is easy to confuse. Newtonian physics can predict a planetary event thousands of years in the future, but there is no Newtonian physics of the mind.

Many books have been written about the science of the mind, love, fate, friendship, story-telling, even religious experience, but so long as we remember that we can't even be sure what thoughts we might have tomorrow, we'll understand that the 'knowledge' that such books give us has to span both physics and metaphysics. If we limit ourselves to 'knowledge' that comes only from our observations or our ratiocinations, our inner life will be barer than Mother Hubbard's cupboard.

We don't need to read a book on neuroscience to grasp the existential truth that we are not just molecules bumping up against each other. We realise instinctively that we are complex tangles of thoughts, feelings, beliefs, longings, hopes and desires, tied with knots that are difficult to unpick.

Telling the whole story

This is because science alone, no matter how reasoned the causes it finds, cannot 'explain' us to ourselves. It tells only a part of our human story. Equally important are the social, cultural, moral, spiritual and political realms which provide reasons and causes of their own.

We need only to watch a typical news bulletin to realise that science, while able to do much to make our lives easier and longer, cannot alone make us happier, more just or less antagonistic towards each other. Biochemistry, genes and evolutionary algorithms, while being necessary, are not alone sufficient explanations of complex human behaviour. For better or worse, our socially formed beliefs 'cause' what we do as certainly as alcohol 'makes' us drive badly or slur our speech.

To get to the bottom of this conundrum of 'what makes us tick' we might set out with good intentions of 'doing' the explicit science first to discover how far empirical enquiry or logic can take us. It is not long however before we come up against aspects of our humanness that only our implicit mind can 'do'.

17

We Are More Than Our Brains

We're not detached knowers or bodiless subjects gazing down coldly on a catalogue of fixed facts. We are flesh and blood seekers thrown into the midst of dynamic change, our senses tingling with constant stimulus. This means that, even with our best science, we never 'see' a single, brute reality. We 'see' nature and ourselves through the intimacy of our perceptions, language, customs, memories, and the theories we have espoused.

The eighteenth century philosopher Immanuel Kant concluded that mind cannot be taken out of the equation of how we crunch sense data to arrive at the 'truth' of anything, even when we gaze down a microscope. Mind is constitutive of how we see what is under our gaze. Incoming information is meaningless without the inbuilt filters and biases of the mind, which have undergone the same selection pressures that sculpted the shape of our nose.

This does not mean that mind is more fundamental than matter, or that reality is just a giant idea in the head of some remote deity, or that mind is outside what can be known. It does mean that our mind is intrinsic to how we understand our brain.

We can't dismiss philosophy as a distraction while we try to get at the physical reality of our brain and the science behind it. We need a 'meta' discipline beyond brain science called philosophy of mind. We have to invite philosophy into the debate if we are to negotiate the social realities of culture, laws, beliefs, political systems and personal prejudices which frame our discourse as palpably as our neurons or synapses.

We have to 'do' the metaphysics as well as the physics, because there are metaphysical realities that are essential if we are to have a sense of self, respect for the person, practical reason, scientific method, liberal politics and meaningful freedoms. As the mediaeval friar and theologian Thomas Aquinas remarked, the wise physician studies the mind as well as the body.

For some this raises the spectre of dualism, pitting knowable facts about the body against airy speculation about the mind. René Descartes regarded his bold claim 'I think, therefore I am' four hundred years ago as proof that he possessed a mind. Since then, some have accused him of buying this victory at the expense of stranding his mind in metaphysical no-man's land, while others have charged him with being deluded that he could put so much trust in the musings of his imagination.

We Are More Than Our Brains

All that mattered to Descartes is that he could prove he really existed. He was a free self, and his thought was his own, not just the product of a mechanical brain. His radical scepticism consumed everything except the fact that he was having this thought.

He concluded that the thinking self is not an illusion, but the starting point of all our knowledge. It is not a case of choosing dualism or rejecting it, which is just another binary pitfall. The combined history and geography of our mind-brain means we are stuck with it. The good news is that dualism comes so naturally to us that as a rule we just get on with life without noticing the join.

Choosing our narrative carefully

Aquinas and Descartes had no interest in setting up battles over materialism, reductionism, determinism, rationalism or dualism. These are just thinking tools, not explanations of how the world works. They knew that we need reasons to get out of bed in the morning other than to cram our mouth with calories for another day painting illuminated manuscripts, or speculating how the mind works.

They understood there is no gap between the mind and the world, because our brain is not just a coding device or information processor. It is an active creator of the reality we inhabit. We are implicated participants in our life-world, and as we choose, so we become. Every decision we make changes our brain a little.

They also understood there is no single thing called 'science', but a complex of disciplines that contribute differently to our understanding of the whole, or of matter and its relation to mind. This explains why, once science impacts on our life choices, it is no longer 'pure', but laden with the concerns and prejudices of the day.

We see this clearly in the various ways 'evolutionism' has been interpreted through shifting metaphors since Darwin: as an arms race or mutual aid, blind chance or discernible purpose, random drift or a balanced tree of life, ruthless weeding out of the unfit or inclusive adaptation, steady progress or sudden leaps, the triumph of self interest or the vindication of cooperation. Post-Darwinian 'selection' theory at one point became entangled with ideologies of historical inevitability, eugenics, ethnic cleansing and sterilisation of the 'feeble minded'.

Neuroscience has been caught in similar ideological wars. Is the mind reducible to the brain, and the information it carries? Is it a sexually selected entertainment system, an evolutionary fudge, a Machiavellian schemer, an over-blown computer? Should we reverse-engineer our brain to the point that human intelligence can be gifted to machines that will eventually outsmart us? Should we go along with the desires of some transhumanists who want to enhance their brain to the point that they become trans-species cyborgs that could not evolve by natural means?

Our new-found knowledge of the brain needs to be seen for what it is, exciting but far from comprehensive, and certainly not a basis for social policy. Nature does not speak for itself, but responds to the questions we ask of it, and how we apply the power that our new knowledge gives us. This is always in the context of the chapter that has just been written, or the next chapter we want to write in our story. No science is ever simply value-free knowledge or research. It is always part of a more widely negotiated agenda about the direction we wish our society to go in.

As we proceed further into the Age of Reverse-engineered Brains, Artificial Minds and Machine Intelligence, we must remain in control of the narrative, because each twist in the plot shapes our understanding. By insinuating itself into our consciousness, AI creates a new normal, influencing how we see ourselves. On one hand, we need not be too concerned, because mind is an irreducible aspect of our experience, not explainable by brain function alone. It is the storehouse of generations of spiritual wisdom.

On the other hand, we are compulsive storytellers, given to telling different stories about ourselves, as we see this in the 'Looking in from the outside/out from the inside' illustrations at the start of the book. We drift between the 'day language' of studying the mind as a detached object in time and space, and the 'night language' of mind as experience lived from the inside.

We Are More Than Our Brains

Different paths to understanding

This conflict of narratives was present in the very first thought that dawned in the mind of our ancestors roaming the African savannah. How do we integrate the outside theory of science with the inside theatre of lived experience? It's worth noting that both the words theory and theatre go back to a common root to do with ways of seeing.

The laws of nature appeal to us because, like a single-issue politician, they beguile us with clarity and simplicity. The straight lines of gravity must be obeyed. Culture operates to different laws of cause and effect, embroiling us in a much messier pursuit of meanings which unexpectedly curves, even curls back on itself.

For this reason no 'laws of the mind' have ever been established, except perhaps as something called the perennial philosophy, or the accumulated wisdom of mankind. This shows remarkable consistency of thought and value, such as the Golden Rule, though its testing ground is the cut and thrust of everyday life, not the empirical observation in the laboratory.

Outside of the Golden Rule, or a general agreement that mutual aid wins in the long run, there has been a bewildering flurry of 'isms' in philosophy, religion, art and psychology in all known eras and cultures, each an attempt to deconstruct how our mind processes reality, and what it means to be human.

Each 'ism' is a product of the questing human mind, seeing through the prism of its own culture, responding to the knowledge available in its day. This is as true for science as it is for philosophy. Alchemy is now seen as amusing wizardry, but it laid the foundations of modern chemistry.

Neuroscience has its own pedigree of 'isms'. It is tempting to see it as a single study, but it has many tentacles reaching into related disciplines of cognitive science, social psychology, molecular genetics, evolutionary psychology and information theory, to name only a few. In its brief history it has toyed with atomism, associationism, phrenology, behaviourism, materialism, determinism and reductionism. Only in the last couple of decades has it come out of the laboratory to consider how none of these 'isms' adequately explains what happens when brains interact with each other in human groups, which is the reason they evolved in the first place.

We Are More Than Our Brains

In philosophy, there are even more 'isms', and 'truth' has even more faces. Each 'ism' is a staging post, giving us only a snapshot of reality, because human thought and enquiry never arrive at a final destination. Nor does it make sense to dismiss all previous attempts at understanding as 'wrong turns' that have delayed our journey to present omniscience. What we know today is built on what we knew then, and what may still turn out to be true or useful.

Keeping our eye on the whole

Modern neuroscience sits at the cutting edge of a long exploration of mind-body dualism in Western thought. Some see this philosophical quest as a glorious failure, leaving the brain tied to its body and the mind cut loose with nowhere to go. Galileo got round this impossible split by excluding mind completely from his physics. Forget the philosophy, and focus on the speed of objects falling from the Leaning Tower of Pisa. Everything is reducible to numbers.

This exclusion of metaphysical speculation about things unseen set science free to take nature apart for four hundred years, resulting in the technological dominance we enjoy today. Analysis and reduction have proved to be powerful tools in cracking open the secrets of things 'out there'.

But there has been a heavy price to pay for these material gains. Our concept of what passes as 'real' has been shrunk, we no longer feel connected to nature, we've lost our sense of the world as flow, and what goes on 'in here' has been relegated to the touchy-feely zone.

The irony is that, even as our scientific knowledge and technological mastery have forged ahead, personal happiness levels, social harmony and the health of the planet are in retreat. We do not flourish well when we detach our brain from its body. They are bound together with hoops of steel, if only because a disembodied mind is an evolutionary impossibility.

It is not surprising therefore that philosophers have felt drawn in the last hundred years or so out of their ivory towers into the laboratory and computer workshop. What light do cognitive science and AI research throw on our understanding of reality, our concept of truth and what it means to be human?

We Are More Than Our Brains

Scientists have also approached the mind from a variety of angles, exploring by turns its origins, structures, functions and failings. There has been a growing realisation that the 'meaning' of the mind they seek to explain resides in the interconnectedness of their respective disciplines. They can't grasp the whole by minutely analysing all the parts. However precisely they describe a brain as so many interacting neural operations, they are still left with a person in their laboratory wondering what to have for lunch.

A brain is not reducible to molecules, information or mathematics, but always a unified network of inputs and outputs, events and experiences, belonging to an organism inhabiting a specific life-world, the neurochemistry of its brain as hard to unravel as the nuclear physics at the heart of a supernova.

Scientists try therefore to keep their eye on the big picture, as well as on the component parts. Physicists look for unified force fields, chemists seek integrated reactions, biologists search for interdependent ecosystems. Nothing exists in isolation. Niels Bohr, whom we have already quoted, advocated the principle of complementarity: the mind complements matter just as particles complement waves. This broader understanding allows us to see that the non-material realms of mind and reason are as 'real' as material sticks and stones.

This thinking applies to all theatres of human enterprise, though some cherry pick their reasons. Free marketers gladly trumpet the economic theories of the eighteenth century thinker Adam Smith to justify egotistical self-advancing capitalism, but they conveniently overlook the fact that he wrote just as much about the importance of altruistic sentiments, or the social bonds that make an economic community possible in the first place. There is no self except as a member of a wider society.

By looking inside the brain, at what happens when we engage socially with each other, neuroscience has confirmed the core beliefs of developmental psychology, and what our grandmother regularly found her own way of telling us: our mind-brain is a complex cognitive-emotional whole that goes from the crown of our head to the tips of our toes. Our brain spots threats and processes inputs, but without feelings and culture that connect us to each other we are, in Shakespeare's words, 'poor, bare, forked animals', with no accommodation to call our own.

23

We Are More Than Our Brains

No view from nowhere

Scientists do their best to show us the physical world as it really is, but there is no objective view from nowhere. There are only interpretations of perceptions, from which flow a metaphysical reality. It's easy to forget that all our concepts, though dependent upon our percepts, are constructs at one remove from them. We see through the window of the time we live in, making us not perspectiveless gazers but situated observers.

In addition, although science tells its story in numbers, which promise a kind of certainty, our tentative journey into mind is recounted by words, which are full of ambiguities. Objecting to the inadequacy of neurospeak is not therefore mere sophistry or the saving of appearances. We must watch our language, or at least our metaphors, which is the point of having a free-thinking and self-critical mind in the first place.

To say for instance that we *are* our brains suggests an equivalence of identity, but this is a misleading verbal trick. When we say coal *is* black and Romeo *is* in love with Juliet, we are playing different word games. The brain *is* no more the mind than the communion wine *is* the blood of Christ. We are more than our brains, because the material ingredients of the brain can't alone account for the immaterial thought that we *are* more than our neurochemistry or genes. We might as well claim that we *are* the people we love, or who love us in return.

The mind sees more than the eye, as we realise every time we look at someone we love. This means there is always a realm of values and goals beyond materialism and determinism, which physical process and natural science can neither generate, prove nor disprove. Above and beyond the hormones buzzing through our thalamus, we choose who we love, we decide who to be kind to, and the only backstop for such beliefs is our convictions, which need constantly to be put to the test.

Our every thought transcends the atoms that comprise it, inseparable from its moment in time. We are never disembodied nobodies, but always somebodies reacting to someone or something somewhere. This means that we never have a single view from anywhere, or even nowhere, certainly not everywhere, but always a multiplicity of views from manywheres.

We Are More Than Our Brains

Taking the mystery to pieces

Given these complexities, some neurophilosophers (as distinct from neuroscientists, who tend to stick to the data) have opted for eliminative materialism: to be fully known, the existential mind must be pared back to the biological brain. This sleight of hand can be achieved by making the mind *sound* quasi-physical: it's a phase transition, supervenience, strong emergence, computational architecture, quantum effect, complexity from simplicity, or smart outcomes from stupid inputs.

These metaphors merely replace one mystery with another. Also they muddle levels of explanation. They give no idea how neurons make knowledge, axons cause anxiety or dendrites make decisions. Reductionism is useful for establishing objective downward explanations and the naming of parts. A clock for instance can be stripped to its tiniest components and then reassembled. If we do this properly, it will be able once more to tell us the exact time.

Humpty Dumpty
Because the arrow of time is irreversible, an egg can't be put back together again after it has been broken or taken apart. What happens when we crack the mind into a frying pan?

A cloud presents a very different kind of challenge, not static but dynamic. Like Humpty Dumpty, it can't be taken to pieces and put back together again. We see a cloud floating across the sky,

25

but what we don't see inside it is its energy patterns, colliding particles and mathematical complexity. Also, unlike a clock, a cloud is transient, so we can never gaze at the same cloud twice.

The wonder of a cloud reflected in a lake is an even greater mystery. As with most things in life, we don't need to strain for reasons or explanations. Instead we accept that the cloud does not ask the water to reflect it, nor does the water owe the cloud anything. Sometimes things just are the way they are, happening serendipitously inside and alongside each other. Clouds, winds, storms, ocean currents, sunny days and friendships are 'made' at the microlevel, but they 'happen' at the macrolevel.

Like the mind, a cloud is a hierarchy of causes, constantly on the move, making formations that will never be repeated. This is why weather forecasters often get things wrong. Weather fronts and tornados cannot be reduced to brute matter or inexorable cause-and-effect, because they are complex systems with too many moving parts and unpredictable outcomes, causal but also casual.

Brains are also like this, dynamic, global and complex systems, as any neurologist will testify, hardly reducible to the linear or logical. Over-zealous reductionism, stripping the system down to atoms and squeezing out the 'I' of the owner, is not therefore the right tool to reveal the intersubjective richness of experience, account for our relations with each other, or give us reasons to believe. As the physicist Freeman Dyson remarked, we are human beings first and scientists second.

By reducing all qualities to quantities, we merely deepen the mystery. Where the mind is concerned, what we need is not reductionism, but holism. The mind-brain is an open field criss-crossed by all the paths we have ever travelled, and roads not yet taken.

A brave new story
This does not mean our mind can roam free of our brain. We cannot choose our parents, or the place of our birth. We cannot escape our double inheritance of the complex intelligence of a big-brained primate, hitched to the powerful emotions of a socially needy mammal. We cannot dismiss our human tendency to sway between anger and compassion, hope and despair, smartness and

stupidity. We can't deny that we can feel spiritual uplift one moment and then be floored by a virus the next.

In this book we explore how this duality plays out through the new findings of neuroscience, sociobiology, evolutionary psychology, information theory, behavioural economics and cognitive science. Can they give us a brave new story, or a master theory of how we get the wine of consciousness from the water of neurons? Can they help us become our own soul-doctors? How do we derive human nature from a genome barely different from a chimpanzee's?

It seems that with every discovery about the brain and the genome, the complexities of the self, free will, imagination and creativity do not come closer but slip further from our grasp, along with the perplexities of depression, schizophrenia, autism and common unhappiness.

And yet there have been enormous gains. Neuroscience has transformed medicine, psychology and psychiatry. We now have much a richer understanding of brain disorders, finer neurosurgical techniques, more effective targeting of our synapses with drugs, and a deeper grasp of the processes of addiction.

Cognitive neuroscientists have exposed our bad thinking habits, and shown how we can correct them. Evolutionary psychologists have made us more aware of how our mind works, and the kinds of minds that other animals possess. Sociobiologists have given us clues to the puzzle of why we often behave as if men and women come from different planets. Behavioural economists have shown us why we so often back the wrong horse.

Philosophers of mind have joined the fight by helping to explode misleading or damaging myths about the brain. IQ is not determined at birth, we are not naturally selfish, madness is not caused by demon possession, emotion is not inferior to reason, our brain doesn't start to die the moment we are born. Addicts can be rehabilitated, criminals can be reformed, women can read maps. Depression, autism and schizophrenia are conditions of mind or states of consciousness, not divine curses or moral failings.

The jury is still out on whether free will, self and consciousness can be reduced to algorithms, but these debates pass largely over our heads. The 'discovery' of neurodiversity however, which is neuroscience's version of a much older teaching of the brotherhood of man, is of a different order. It has major implications for why we should value the differences between us, because they emanate from a biology and brain configuration that we all share.

Beneath the skull we all work to the same neural hardware, but beneath the skin we are all different. There is no biological foundation for racism, misogyny, homophobia and xenophobia. These are cultural prejudices which are out of place in a globally connected world of seven billion people charged with sharing a fragile planet. If neuroscience can help us banish our ancestral gremlins, we need as much of it as we can get.

We must beware however of replacing old myths with new ones which can be just as imprisoning. There is no 'God spot' in our cranium, and religious faith cannot be reduced to neurochemistry. The only way to understand what Moses was seeing and feeling during his vision on Mount Sinai is to immerse ourselves in his mindset, his mission and the people he felt answerable to. Whether delivered from on high or as a distillation of the law that helped his nation emerge from idolatry, the Ten Commandments still stand as the founding principle of Western morality that we take for granted today in its secularised form.

Properly interpreted and kept in proportion, neuroscience is a great advance that banishes ignorance, giving us the basis for a new understanding of mind, or 'what makes us tick'. If we are also to become fulfilled persons, mindful citizens and meaningful choosers, we need a bigger story that integrates our biology with our culture.

Our mind-brain means nothing except as lived experience, as an open-ended narrative that we write. Taking our brain apart helps us to fix it when it breaks down. To be good users of a healthy brain however, we also need a deeper understanding of our mind, enabling us to unify thought and feeling in causes greater than ourselves. We need to avoid the either/or dualism that has blighted so much of the debate about the type of people we have it in us to be.

Mr Popper's Three Worlds
The philosopher of science Karl Popper offered us a way to reconcile our brave new knowledge of the brain with our long-

cherished theory of how our mind works. He suggested that we inhabit a trinity of worlds, distinct but indivisible.

World One is the realm of physical objects, biological organisms and knowable facts. Popper knew that science thrives on certainty in a world of things, amassing data and refining theories to account for how 'stuff' goes about its business in predictable ways. It is because matter behaves in these consistent ways, even in our brain, that neuroscientists are able to trace the patterns of our neural activity on an MRI scan, perhaps one day finding a treatment for schizophrenia or a way of preventing Alzheimer's disease.

World Two is life as it appears to us through our senses, manifested in experience. This explains how each of us can 'see' the world slightly differently, or vote for a different political party. This does not mean that relativism rules the roost, or that values are a free-for-all. When we travel the world, we find common courtesies, shared emotions, universal logic and consistent recognition of colours, albeit with slight regional variations.

Sir Karl Popper
1902–1994
Popper divided our knowing into three worlds. Our scientific knowledge of the brain belongs to World One, our lived experience of the mind belongs to World Two, and our cultural being to World Three.

In other words, though biology makes a brain, experience makes a mind. These are equal truths, both essential to understanding the human condition, both revealing a world of wonders. Science can reveal to us the laws of matter, but not the rules of life. Neuroscience can explain what is going on beneath the waves, but only the skipper up top can steer the ship.

29

We Are More Than Our Brains

Popper's Worlds One and Two reframe the ancient mind/body dualism that has for so long dogged discussions about the mind-brain. There is a ship, and there is a captain, but we need to understand how they work together to make a voyage. Descartes was right. We think, therefore we are. There is thought, and it thinks us. Popper suggested we best understand this conundrum of reciprocity through World Three, which not only integrates Worlds One and Two, but also creates an independent reality above and beyond it.

The human life-world

World Three is not an airy-fairy 'other-worldly' realm, because it has human fingerprints all over it. It is our life-world, comprising all the things we do together as a species, and all the institutions we have built, such as language, culture, science, religion, music and law. It is all the ideas that anyone has ever thought, especially those that have caught on to make lasting social realities, such as the groups we belong to, political systems, scientific theories, artistic traditions and technological achievements.

Every time we participate in Mr Popper's World Three, something important happens, affecting all aspects of our being. The world changes us, and we change it. In other words, when we act in the world, we are also acting on ourselves. We make and are made in equal measure.

Go back to travelling around the world for a moment. If we break the law in a foreign country, we find ourselves confronted with the weight of a system that is founded on very similar principles to our own: the wrongdoer must make restitution for wrongs committed. Popper rightly claims therefore that World Three is just as 'real' as Worlds One and Two. When we're in prison, it matters little whether it's the walls, prison officers or local byelaws that are holding us captive. All we know is that we can't escape.

It's in World Three that our science gradually impacts on our ideas and beliefs, first shifting our attitudes and then changing our social practices, around gender identity, drug legislation, treatment of mental illness, stewardship of the planet, education of the young, care of the elderly, global cooperation, our eternal pursuit of happiness.

It is impossible therefore to study the brain in a cultural vacuum. In World Three, like the sea, land and sky, everything is whole, embodied and connected. So although we sometimes need to split things up in order to study them, we must also remember how to fit them together again.

The pursuit of meaning
The problem is, World Three is expressed in language, not data, and words are notoriously imprecise, slipping and sliding away. We have evolved as compulsive meaning-seekers, but we cannot come up with any repeatable laws about thorny social problems, only faltering attempts to solve them in the light of present knowledge, which is never complete or perfect.

In World Three, meaning is always provisional, unlikely to yield to bullish reductionism. It cannot be generated from within the system, by genes, neurons and algorithms. The meaning is not there until *we* make it, looking down on the system, as individuals and groups acting in a common cause.

The same can be said about our values, which emanate from the way we solve our problems as they arise. If we value truth, health, beauty, justice and goodness, it is because we have proven that these qualities work in the interests of all. This is what the Ancient Greeks meant by the Logos, an overarching principle that marries physical necessity with moral law. This core value and universal truth is as evolved and consequential as genes, neurons and the forces that draw two lovers together.

'We are our brains' is therefore only a temporary and partial insight, the product of a particular and limited method of enquiry. Given that we constantly change our views, and we remake ourselves in the light of new understanding, we need to remember the opposite but complementary truth. We are more than our brains. We are our minds.

In fact, by a strange quirk of the evolution of human consciousness, we might argue that we were minds *before* we were brains. Centuries ago on the African savannah, we felt more in tune with the movements of the stars above than with the mechanics of the biological organ just behind our eyes. Our first

31

question therefore is: why did it take us so long to realise we possess a mind-brain, and to fathom its secrets?

We are going on a mind hunt, starting with the extraordinary origins and achievements of neuroscience, and how it has thrown light on the recesses of our mind. Where did the 'idea' of the brain come from, and what do we learn from its history? We review what a scientific adventure it has been to 'discover' our brain, and what a philosophical puzzle it has been to get inside our mind. We reflect on the fact that the human brain did not mysteriously appear fully-made, but is the serendipitous product of eight million of years of primate evolution, and all the gremlins that go with it.

These very defects are the driving force behind our greatest achievements, such as our curiosity about our own mind, our ability to love each other beyond the call of duty, our passion for creating art, our yearning to formulate belief systems, and our confidence that we can invent a better future.

In the second chapter, having explored our brain's history, we investigate how its geography shapes our thoughts. We journey through the secret chambers of the mind, and consider what light neuroscience has shone into them. How do we make one mind from two brains?

Neuroscience has brought us a long way in its first hundred years, but in some regards, with its present methodology, it has told us all it can. To take us any further, we don't need more and more answers about the same things. We need new questions about ourselves, starting from the premise 'We are more than our brains'.

Discovering the brain's history

How has our brain kept its secrets from us?

Phineas Gage - phrenology - early brain science – Thomas Willis - modern neuroscience - neurophilosophy - making new neurons

- Early thinking about the brain focussed on the heart as the seat of reason and controller of the body.
- Until quite recent times, understanding what lay inside the brain was limited to what was visible on the outside.
- The accident that befell Phineas Gage in 1848 marked a dramatic shift in realising that the brain is the controller of the body, and the shaper of our personality.
- Key breakthroughs were the discovery that electricity powers the nervous system, followed by the 'seeing' of the neuron, the synapse and the neurotransmitter.
- Neuroscientists stick to the data, but neurophilosophers are given to speculating about the foundations of self, consciousness and free will.
- By eliminating the 'ghost in the machine' however, they undermine their own credibility, like the tree surgeon who saws off the branch he is sitting on.
- In contrast, the evidence-based discoveries of neuroplasticity and neurogenesis have transformed psychology and psychiatry.

Meet Phineas Gage

Our modern understanding of the brain is surprisingly recent, perhaps going back no further than a bizarre industrial accident in 1848 to an American railroad construction worker, Phineas Gage.

33

We Are More Than Our Brains

He was overseeing the use of dynamite, and one day made a serious error: he detonated the fuse prematurely, and a metre-long tamping iron as thick as a broom handle was shot up his nose, going clean through his right frontal cortex and out of his skull.

He survived the explosion and lesion to his brain, seemingly recovering within hours. He could still walk and talk, but in the following weeks his fellow workers noticed two key changes. His personality had changed, and he had lost any sense of planning for the future.

He lived another thirteen years, but they were not happy. His behaviour became erratic, he uttered profanities, lost his popularity, suffered from emotional instability, and ended up as a fairground freak, eventually dying from an epileptic fit. But for this traumatic experience, his name would be lost to history. Instead, he has become the poster-boy for neuroscience, and has even had a pop group named after him.

His name lives on because his accident, and his life after it, helped to break centuries of religious taboo and medical ignorance about the brain. Religious dogma and common humanity had forbidden any investigation into live brains, but Phineas was a living experiment.

His partial brain injury demonstrated beyond doubt that the brain is controller of the body and mind. More specifically, Phineas's left hemisphere was still intact, so he could control his body functions and movement. His right was damaged, so his thought and personality were affected. Phineas's ability to live a kind of life after the accident demonstrated that, if one of the frontal lobes is injured, the other can deputise up to a point, the extent of the lesion determining how much function is lost overall.

Since Phineas's death, neuroscientists have unravelled many mysteries of the mind that used to be put down to insanity, demon possession or weak character. The injury to his brain was plain to see, but not all brain lesions are visible from the outside. Deep inside the brain, a stroke, burst artery or tumour pressing on key areas can result in sudden change of personality, sex addiction, hallucinations, even a sense of feeling dead in the midst of life.

We all have a fleeting glimpse through synaesthesia of what can happen when the wiring and economy of our brain is atypical or disrupted: we might see numbers as colours, or hear tastes as sounds. This is a fairly harmless experience, possibly a pleasant one, but Phineas's accident opens a window onto congenital malfunction and accidental trauma to the brain in conditions as diverse as Tourette's syndrome, multiple personality disorder, obsessive compulsion and schizophrenia.

34

His injury reveals an even deeper paradox, one that is testament to the brain's versatility. Although a 'local' part can be deleted, other parts maintain a 'global' operation, so life can carry on, albeit radically altered. The brain is a curious mix of fixity and flexibility. Some of it is prewired, but it retains a vital capacity to rewire itself, especially when we are young. It says a great deal for Phineas that, despite his horrific brain injury, he managed to do other jobs and enjoy some quality of life.

This sounds a note of caution to those who seek to pinpoint particular spots in the brain where love, sexual desire, psychopathic tendencies or religious experience might be located. We cannot say that Phineas's loss of part of his cortex 'caused' his personality change, or that if we remove a particular part, we can create an obedient zombie or cure depression. All we can say is that each part of our brain contributes to a complex whole, and when we eradicate or lose millions of neurons in a single area, even the most skilled neurosurgeon cannot safely predict how our behaviour will be affected, and with it our sense of who we are.

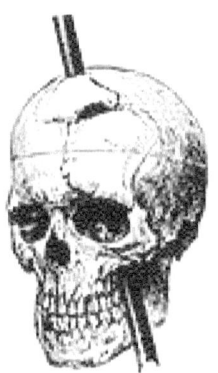

Phineas Gage's Skull
This shows the entry and exit point of the tamping iron in Phineas Gage's skull after his death in 1860, and the extent of damage to his frontal cortex, resulting in dramatic personality change and loss of ability to plan the future.

Early brain science

Speculation about the workings of the human brain had been rare up to the point of Phineas's accident. The mind was the domain of the gods, and it was not appropriate for mere mortals to probe too deeply. In any case, it was assumed that the spring of human behaviour, or the soul, was to be found in the heart, not the head. One of Shakespeare's characters ponders the question: 'Where is fancy bred, in the heart or in the head?' There had long been intimations that it was the latter.

We Are More Than Our Brains

In a very early acknowledgement of the brain's physiology five thousand years ago, an Ancient Egyptian physician made a note on papyrus of specific wounds to the brain sustained by soldiers in battle, and how their subsequent behaviour was affected. When the Egyptians entombed their dead however, they carefully dried out the liver, lungs and intestines and placed them in canoptic jars beside the mummified body, as if they somehow contained the essence of the deceased. The brain was sucked out of the skull through the nose, and discarded.

The Story of Brain Science	
The ancient world	Thinking and the soul reside in the heart. The brain is generally considered to be a mysterious mush of mucilage and ectoplasm.
The long 'dark ages'	Religious leaders largely forbid dissection of human bodies, though one or two speculate that the soul may be located in the ventricles or cavities in the brain.
1500's	The Flemish anatomist Andreas Vesalius publishes his remarkably accurate woodcuts of the brain in 1543, but the mind is still stranded somewhere between the heart and the head.
1600's	René Descartes introduces dualism by suggesting that mind and body are two substances, with only a tentative guess how they are connected in the brain. Thomas Willis conducts what can be regarded as the first 'neuroscientific' investigations.
1700's	There is increasing belief that the brain is the organ of thought, but great puzzlement how it gives rise to mind.
1800's	The materialist view gains strength that everything that appears to the mind is attributable to what happens in the brain.
1900's	The neuron, the synapse and neural networks are discovered. There is no longer a need for a ghost in the machine. The phenomenon of the mind is reducible to the physics of the brain.
1950's onwards	The brain is a computer running on information. The first 'machine minds' appear.
Modern neuroscience	Some neuroscientists are convinced that we will soon be able to reduce joy, sorrow, schizophrenia, depression, consciousness, mind and free will to 'neural correlates'. Others believe that neuroscience needs to start asking different questions about how mind and brain interconnect.

We Are More Than Our Brains

Writing centuries later in a Mediterranean culture, the 'father of medicine' Hippocrates took a very different view: the brain is the seat of the soul, not our internal organs. 'From nothing else but the brain come joys, sorrows, delights, laughter and sports'.

His fellow Greek Aristotle was not persuaded, insisting that the job of the brain is to cool the blood. Thinking takes place in the heart, a sentiment echoed in our insistence that we believe something with all our heart. He also believed that the head is round because it mirrors the perfection of a circle, but this is a case of putting theory before the fact. Nevertheless, his notion of a heart-mind partnership gets us as close to human experience as the modern pairing of a mind-brain, both concerned to reconcile thought with feeling.

In the second century the Greek physician Galen looked not to the gods but to anatomy for an explanation of how the body works, but dissection was considered a violation of the soul, so little progress was made in understanding the brain. Instead, the theory of humours prevailed for over a thousand years, based on the idea of fluids in the body which keep the brain dry, moist, hot or cool.

This was more a kind of elementary psychology of mood and personality, not brain science. It was to be a long wait until the sixteenth century, when the Flemish anatomist Vesalius published a book of drawings of the brain so accurate that they are still consulted today.

Thomas Willis

Many date the beginnings of modern brain science to the seventeenth-century English physician Thomas Willis who, in an age barely emerging from witch-hunting, dared to peel back the skull of a live dog to observe for the first time the flow of blood through the brain. It is highly unlikely that the dog survived the experiment. Willis also charted the tracery of the nerves connecting the brain to its body, calling the spinal cord 'the King's highway', and coining the term 'neurology'.

Willis is almost forgotten today, but he was a pioneer in realising that the brain is more than 'a bowl of curds', as one of his contemporaries dismissed it. His search for the 'spirits' that motivated the mind laid the foundations of modern neuroscience. He knew that messages are carried along nerves, not ferried around the body in 'humours' as taught by the Ancients, and that the brain sits at the apex of a connected network. He had no way however of knowing what animates or powers the machine.

37

Like many physicians in the seventeenth century, he found it difficult to break free from bleeding and purging the patient, but he was also determined to establish natural causes with effective treatments, dispensing with alchemy and superstition. He found it difficult to say farewell to the notion of a soul, but he could not identify the 'vital energy' that connects the mind and the body, leaving him stranded in a dualistic split between a material body and a non-material mind.

It was to be another hundred years before Luigi Galvani demystified the power source of the nervous system by running a current of what he called 'animal electricity' through a newly hanged corpse, causing it to jump. He had discovered the vital force that powers life and animates matter, connecting the physical brain to the metaphysical mind, inspiring Mary Shelley to write her novel 'Frankenstein' in 1815. He also anticipated modern defibrillating machines or heartbeat-revivers by over two centuries.

Willis and Galvani were like Darwin, early pioneers who pointed the way for those who followed. Darwin knew that biological information gets passed on, but could not explain how, leaving it to later researchers to discover what came to be called genes. Even then there was a long wait before the language of the genes was unravelled as coded strings of DNA. Science often advances in this faltering, piecemeal way.

Darwin was working within the new 'mechanical philosophy' that Willis helped to found in the seventeenth century, which included William Harvey's discovery of the circulation of blood in 1628. The secrets of the body and its brain were beginning to yield to scientific method.

Darwin had no place or need for God in such a scheme, but Willis, like many of his age, was not an out-and-out materialist. He understood that the nerves can account for the 'sensible soul' of our body, but to his dying day, and like so many before or since, he could not solve the puzzle of the 'rational soul' of the mind. He would have found some sympathy with the title of this book: we are more than our brains.

The thinking of the doctors who attended Phineas after his accident would have been dominated by a now discredited science called phrenology, which relied on bumps on the skull to indicate what lay beneath. This was barely an advance on the mediaeval lore of physiognomy, or judging character by facial features. A big nose was a sign of lust, and eyes that were too close together signalled someone who could not be trusted.

We Are More Than Our Brains

In 1809 Franz Josef Gall had moved beyond physiognomy, but not phrenology. He looked for telltale signs in the shape of the head, convinced that each personality trait had an equivalent surface protrusion.

This led to a practice of identifying criminals by giveaway head-lumps indicating vice or villainy, amativeness or combativeness. To this day a cast of the outlaw Ned Kelly's skull sits proudly on display in the Old Melbourne Gaol. Many doctors used to sport a ceramic head on their desk inscribed like a county-map of England with moral and cognitive boundaries, but it is now universally accepted that there is no correlation between the curve of our lip, the shape of our head, the intelligence of our brain and the worth of our character.

Going inside the brain
If the brain was to yield its secrets, a more scientific and less superstitious approach was needed than phrenology, and this meant looking under the bonnet. In 1861 the French surgeon Paul Broca ushered in a new era in brain science: he pioneered a way of going inside the brain. He nicknamed a patient 'Tan', because that was all he could say after damage to his brain just above his left ear. Broca assumed that this area controls language production, and it is called Broca's area to this day.

Broca's theory comes with a health warning. Neuroscientists have since discovered that Broca's area does play a role in language, but like memory and perception, the brain's language operations are spread across many regions. To confuse matters further, brains have unique configurations. Some people have a healthy Broca's area, but suffer nevertheless from speech and language problems. Some people have a diseased Broca's area, but can still communicate well.

We can't therefore be over-precise in our brain-mapping assumptions, especially when we consider that, in rare instances where neurosurgeons remove a child's left hemisphere completely to prevent life-threatening seizure, the right brain is 'plastic' enough up to the age of ten to deputise for nearly all of the left's activities, including language. We now understand that brain faculties are not confined to precise borders but are linked by tracks and trails across the whole brain, memory for instance traversing many areas simultaneously.

If the geography of the brain is difficult, the mechanics are more so. As early as 1873 Camillo Golgi established that the brain is a structure of neural nets. But what stitches the holes together? Answers started to appear in 1906, a crucial year for brain science.

We Are More Than Our Brains

Sir Charles Sherrington, the pioneer of the nervous system, identified the neuron or master brain cell, having already named the synapse in 1897. He also demonstrated the existence of different nerve pathways for messages to feed into the brain. Paralysed people still feel the full range of emotions even when all their sensory connections to their body parts are cut.

In the same year, Ramon y Cajal discovered a way of staining slides of brain tissue so that he could study through his microscope the dense network of delicate threads between neurons. He speculated, as Darwin did about the existence of genes, that there must be a tiny gap between neurons, even though he could not see it. This was the synapse, across which information is passed, allowing each neuron to operate as an input/output device, receiving a message, generating a response, and distributing it to neighbouring neurons

Ramon y Cajal
1852–1934
Cajal is regarded by many as the founding father of neuroscience. He won the Nobel Prize in 1906 for his pioneering work on the neuron and synapse.

He received a Nobel Prize for this crucial insight, but it took another twenty years after his death for more powerful electron microscopes to prove him right: complex chemical neurotransmitters 'fire' information through the wet-and-dry mechanism of electrochemistry. The traffic between the tiny gaps between our billions of neurons is what orchestrates the growth of our body, the experience of our consciousness, the waxing of our appetites and the waning of our moods. Slowly the brain beneath the skull was giving up its secrets, and a new science was taking shape: neuroscience.

We Are More Than Our Brains

News of the neuron

The word neurology was coined in the seventeenth century, based on *neuron*, the Greek for nerve, sinew or cord. As a prefix, *neuro* has spawned a whole new language and technology.

Neurosurgeons have pioneered life-saving operations for those with neurodegenerative brain disease. From vague medical diagnoses of neuralgia, neurasthenia and neuritis, neuropathologists detect neuromas, neuroblastomas, neurotoxins and threats to our neuroimmunity. In the surgery, neuromodulation, or the placing of magnets on the cranium to stimulate brain activity, has proved useful in treating symptoms of depression and trauma.

In the laboratory, neurobiochemists reveal the intricacies of neurodevelopment, a process which includes neurulation (what guides the growth of a spinal column from a fertilised egg), neurogenesis (the making of new neurons in the brain), neural pruning (the creative loss of many more), and neural theft (the appropriation of neurons left unused).

Meanwhile, cognitive neuroscientists use sophisticated scanners to provide detailed neuromapping of our neuroanatomy and neurophysiology. Their neuroimaging gives us a new illuminated Genesis story: we are the lights that come on and go off in our head. Neurocomputationalists approach the brain as the manipulation of symbols, neuroconnectionists try to trace its pathways and networks, and neuromechanics study the forces that power its motions. Neurolinguists identify the neuroprograms and neurogrammars that scaffold our thoughts and utterances.

Neurophenomenologists take on the hardest job of all, trying to find the elusive neural substrates of mind and consciousness. Where these cannot be found, they turn to neuroemergence to explain how something so inexplicably complex can arise from ingredients so unbelievably simple.

On the basis that our brain is our neurome, they regard our mind as our connectome, joining us up inside and outside through our neural networks. Biologists extend this thinking into a new field called neuroepigenetics, based on the exciting discovery that our DNA is not a fixed life sentence, but responds to our environment through neurofeedback loops. They have given us three great lifelines for the future of our species, which we shall consider in due course: neuroplasticity, neurodiversity and neuroconnectivity.

41

Psychiatrists are attracted to neuropharmacology as a way of targeting drugs more accurately at our synapses. Psychotherapists progress from loose accounts of mental illness to more objective psychoneurotherapy, aimed at connecting neurosis and psychosis with underlying neural signatures.

These technologies save lives and improve the human condition, but a few confront us with dilemmas beyond anything our mind-brain has so far evolved to cope with: the capacity to produce neurodesigned babies, or to create neuroenhanced adults.

Currently the law allows neuroscientists to grow cerebral organoids in the laboratory of up to a million neurons for essential research on drugs or brain disease. These are in essence mini-brains, as would occur in a human embryo. To avoid ethical complications, the organoids are killed before they are four weeks old, long before they can offer insights into how whole brains operate, or interact with other brains.

The neuroscientific quest
We have seen how interest in the brain as a biological organ has grown steadily over the last four hundred years, punctuated by important breakthroughs. Some pinpoint Albert Hoffman's synthesis of LSD in 1938 as the kick start of modern neuroscience, based on neurochemistry. His work showed clearly that synapses are affected by molecules, changing the nature of our conscious experience.

Others suggest that neuroscience originated with cybernetics and information theory a few years later, at the end of the Second World War. There was some excitement at the discovery that neurons in the brain, like digital logic gates, operate as on/off mechanisms or threshold devices.

It soon became clear that the brain's similarity to a computer stops there. Unlike computer software, human wetware is a distributed network, suspended in an electrochemical soup of changing contexts, negotiated meanings, embodied experiences, shifting values and other minds. The mind is a looped unfolding of a personal and social drama, not a linear strip of coded algorithms.

Nevertheless, neuroscience was on its way. It is now so complex and diverse that even specialists struggle to keep abreast of developments in their own rarefied sub-disciplines, which include molecular biology, cognitive psychology, computer science, information theory and artificial intelligence.

They energetically dispute a bewildering range of 'isms' about how the brain works, such as homuncular functionalism, eliminative materialism and neural connectionism, and a host of competing hypotheses such as dynamic scripts, recurrent processing and global workspace theory.

As a crude division of labour, biological neuroscientists focus on how the brain functions, whereas cognitive neuroscientists concentrate on what makes our mental clock tick. In practice however, they are often busy in both fields at once.

All neuroscientists subscribe to a core truth: the brain is not a mush of mucilage, but a highly structured organ with a complex biochemistry. Beyond that, they are driven by different aims. Some are knowledge-seekers, determined to understand the brain for its own sake. Some are mind-healers, using this knowledge to alleviate suffering caused by pathologies of the brain. Some are techno-fixers, dedicated to enhancing the raw matter of the brain into something smarter, or making machines that emulate if not outpace human intelligence.

Whether motivated by biology or psychology, neuroscientists have been much more successful in prising open the brain's secrets than anything Aristotle could manage. They have made important breakthroughs in finding causes for degenerative brain disease, and identifying the symptoms of neurological disorders, such as schizophrenia and autism. They have transformed psychiatry and psychology by providing physical explanations for mental phenomena.

Neurophilosophy

Neurophilosophy presents a challenge of a very different kind: is there any route from a science of the brain to a philosophy of the mind? To understand how the brain generates consciousness, or how it gets from blobs on a scan to insight and intuition, there must be a theory of the observing mind, which can be generated only by the mind doing the observing. This conundrum is a challenge not known to other sciences. Determining what can be measured about the brain requires the measurer to become the measured, which in any other walk of life creates a conflict of interest.

The difficulty for neuroscientists who venture into philosophy of mind is that, by offering what appears to be a theory of human nature, they must stray from empirical evidence into speculation, setting themselves up as secular priests, but without the incense. They have made significant contributions to the Great Conversation about who we think we are, obliging other disciplines to listen to them, but only by straying beyond 'what the science says' in the process. Science doesn't say anything, except offer back answers to questions posed by particular minds.

We Are More Than Our Brains

Science's strength is that it is predictive, but neurophilosophers can no more predict the future or account for the past than Marxists or Freudians. The mind is a natural philosopher, and philosophers *of* the mind have to remember that reflecting on the nature of reality can only be done *by* a mind, which by definition contaminates objective appraisal with a subjective appraiser.

Renaming consciousness as 'emergent autonomous networks' merely replaces one mystery with another, while relabeling mental experiences as brain events sheds little light on how we get from the firings of neurons to an understanding of the habits of the human heart. If we're not careful, neuroscanning becomes a new kind of phrenology, looking for 'bumps' on the inside, not the outside.

There is no ready reckoner to compute flashing lights on a scanner into the thoughts, feelings and memories of an embodied person, though scanning is getting increasingly clever at mind-reading, and models of the brain are getting more sophisticated, far less invasive and dangerous than inserting probes into the brain.

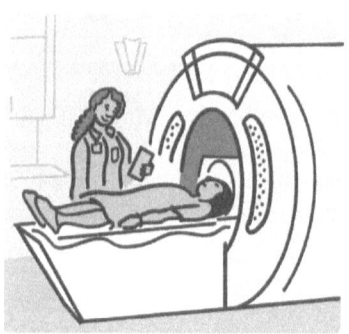

Magnetic resonance imaging
MRI scanners employ powerful magnets to produce a strong magnetic field that forces protons in the body to align with that field. If we can lie still for long enough, and endure the whirring sound, the machine can give a picture of the inside of our brain.

There are many types of scan, tracking blood flow, measuring electrical activity or tracing chemical dispersal, but these offer no way of identifying a saint, sinner or schizophrenic. Some types of scan give structural information, of what happens where in the brain, while others are functional, showing how much activity is taking place after a particular stimulus. Strength of blood flow and oxygen consumption are however merely proxies for what it *feels like* to fall head over heels in love, or how auditory 'sound-bytes' are transfigured into the melody and harmony of music.

We Are More Than Our Brains

Heightened activity in our amygdala when we play a violent video cannot be taken as a cause for increased aggression, only a correlation. The amygdala regulates our whole emotional range, not just our fear, so excising it would turn us into a zombie. Research shows that everything in the brain is a function of up-scaling. Provocation makes us angrier only if we are already heading that way. It might make us more forgiving.

Neuroscience, like all new knowledge, confronts us with moral and existential choices. What do we *do* as a society with our new-found power to eavesdrop on each other's neural activity or thoughts? It also challenges our assumptions about human nature and our sovereignty of mind.

Thomas Willis was aware of the implications for the mind of his new brain science. What becomes of our mind while we are probing our brain? Dualism is an ancient teaching that proposes that our mind somehow exists separately from our body. Most neuroscientists dismiss this idea, saying that studying the mind without investigating the brain is like examining our hand while ignoring our fingers, or trying to understand how a bird flies without looking at the anatomy of its feathers.

The philosopher Gilbert Ryle likened the hunt for the mind to someone standing in a room containing a cooker, dishwasher, sink, food cupboard and pot stand, and asking 'Yes, but where is the kitchen?' Similarly, when we are thinking, remembering, daydreaming, planning and feeling, it makes no sense to ask, 'Yes, but where is my mind?' The mind is what the mind does. There is no ghost in the machine running the show. The ghost *is* the machine, and vice versa, and to suppose otherwise is to make a category mistake.

The mind can't be independent of the brain, because that leaves no way of controlling the traffic between the two. And yet there is a conundrum: why is it that mapping all the brain's networks does not and cannot reveal to us the mind and self as lived experience? How do we get from the processing of isolated inputs and outputs to consciousness as an unbroken flow, one moment constantly giving rise to the next?

The limits of neuroscience
Neuroscience is the child of a rational age that looks to science for knowledge and understanding, recognising that there is still much to discover. Compared to politicians and religious leaders, scientists are much more likely to admit their ignorance, showing far more interest in generating the next set of questions than in providing obfuscating answers to old ones.

We Are More Than Our Brains

The brain is so complex that, even after nearly a century of intense investigation, generating huge amounts of data, neuroscientists still can't fathom what goes on in a fruit fly's brain when it tweaks one of its antennae, not to mention how everything fits together in the infinitely more labyrinthine human brain. It is not their fault when journalists dumb down their complex findings into a simplistic sound-bite, such as 'God spot found in brain'.

The impressive roll call of brain science's achievements speaks for itself, not least its explosion of many neuromyths. It is not the case that 'primitive' brains are less 'evolved' than European, that female brains are less intelligent than male, or that we only use a fraction of our brain at any moment.

But neuroscience, like all sciences, has its limits. Science, regarded by many as our highest form of knowledge, does not alone have the power to save us or reveal ourselves to ourselves, because there is no automatic route from brute scientific fact to negotiated cultural value. The mind-brain is not a static illustration in a biology textbook, but a pulsing organism making a thousand complex choices every second.

Explaining natural phenomena by scientific enquiry is not therefore tantamount to resolving our existential problems. We cannot infer from the lights that glow on a brain scan what experience means to its owner. Buddhist neurodharma is a valiant attempt to ally the findings of neuroscience with its teachings about awareness and mindfulness, but it is no easy step from how things work to what they mean, from what things are made of to how we should put them to use.

Neither neurophysics nor neurophilosophy can alone 'explain' human nature to us; we also need metaphysical reflection. Just as there has been a century and a half of debate allowing us to adjust to the spiritual and moral import of Darwin's revolutionary theory of evolution, so we also need time to find ways of mapping the truths of neuroscience onto human nature.

Some neurophilosophers have used neuroscience to speculate about fields of human experience normally left alone by science, such as religious beliefs, love of the arts, sexual inclinations, spending habits and political affiliations. Neuropsychologists now attempt to link all these things to neurocircuitry in the brain, neurosexologists classify men as Martians and women as Venusians, neuropsychiatrists brand us as neurotypical or neurodivergent, neural Darwinists give us a picture of ourselves as highly evolved apes with a delusory claim to knowledge.

We Are More Than Our Brains

Neurons for everything

If we're not careful, we find ourselves on a pointless quest to find 'neurons for' everything, including neurons for looking for neurons. Our deepest thoughts are reduced to neural hardwiring, and neurojargon enters every walk of life. Neuroscience is the triumph of hard intellectual effort and ingenious experiment, but the step change from the facts of science into the realm of values, or from 'is' to 'ought', is no less easy, and in many ways infinitely more challenging.

Nevertheless, if we believe the hype, neuropolitics can explain how we vote, neurolaw can match each crime with a particular neural signature, neurotheology can give us a neurology of transcendence, neuroethics can show how our brain makes most of our key decisions before 'we' do, neuroeconomics can explain why we buy when we should be selling, neuroaesthetics can present us with a biological account of beauty, neuromusicology can lay bare what happens inside our brain at a rock concert, neurocritical theory can deconstruct the meanings behind what we think we mean.

Neuromarketers can give flashy power-point presentations to retail therapists and behavioural technicians keen to neuronudge us into buying things we don't want, and not buying things we need. Sexual desires can be satisfied on neurodating or neuroporn websites. Social media can allocate us to a neurotribe of 'likers' or 'haters'.

Meanwhile, in Silicon Valley, neurohackers are busy reverse-engineering the brain, writing reams of neural coding, pushing neuromechanics and neurocybernetics to the limit in their mission to neuroengineer a neurocomputer. Transhumanists focus on freezing the heads and bodies of neurophiles, suspending them in eternity until they awake in a brave new neurofuture overseen by a neuromancer.

Overused in this way, neuro becomes a meaningless prefix, a kind of neuromania based on a single neuromantra, in a new place called neurotopia, couched in a new neuronarrative: we are reducible to our brains, and we don't know our own minds. Our mind-brain is an unreliable shape-shifter, half bad dog and half naughty puppy, some of it knowing more than we do, the rest unable to account for the actions it takes. Our mind is 'flat', and we are not half as rational or free as we think we are.

We Are More Than Our Brains

Weaving a new rainbow

The Romantic poet John Keats, who died tragically young at the age of twenty six, worried that the science of his day, with its materialistic explanations for everything, was threatening to 'unweave the rainbow', or remove the mystery from nature.

Early explorers of the natural world using modern scientific enquiry, such as Alexander von Humboldt and Charles Darwin, did not objectify nature, but saw it as an interconnected web or tangled bank, of which we are a part. Neuroscientists, inheritors of this tradition, generally follow suit: the more we discover about the brain, the more miraculous it seems. Brain science has expanded our understanding of the wonders of the mind, just as Newton's experiments with light revealed to us the marvels of the colour spectrum.

We are not reduced, but enriched. We are not made less free, but potentially more rational. We now have no reason to blame homosexuality on moral depravity, schizophrenia on cold mothering, or mental illness on character flaws, because we can shift the focus to the impact of brain chemistry on human behaviour.

As part of a powerful 'zeistgeity' post-Darwinian assessment of our species, neuroscience has teamed up with other disciplines to give us a smart new human narrative, reflected in a steady stream of books which make no mention of 'mind' in their titles: The Naked Ape (1967), The Soul of the Ape (1969),The Articulate Mammal (1976), Neuronal Man (1983), The Ape that Spoke (1990), The Third Chimpanzee (1991), Ape Man (1994), The Moral Animal (1994), The Chosen Primate (1994), The Thinking Ape (1995), The Symbolic Species (1997), The Upright Thinkers (2015) and Sapiens (2015).

Neuroscientists know that, in line with best practice, they must ignore the personal if they are to focus on the physical. They can tell us what happens to our brain when we take drugs, but they leave us to work out for ourselves our psychological need for them, or how as a society to control their abuse. They give us causes for physical events, but not reasons for human choices.

This explains the strained relationship between neuroscience and psychology. Some 'proper' scientists dismiss psychology for merely stating the obvious, without giving reasons or causes. Psychologists criticise neuroscientists for being glued to their machines in the laboratory, producing lots of pretty pictures of brain activity but not accounting for how people actually *behave* in the real world.

We Are More Than Our Brains

How a brain performs in a scanner or at a computer console on limited tasks isn't necessarily replicated by how a mind interacts with other minds in complex social situations. A picture of a brain, whether working flat out or showing lesions, gives no clue about the experiences, choices and feelings that have led to this moment.

Neuroscience is a young discipline, its practitioners readily admitting there is much yet to discover. They know that, if they are to weave a new rainbow, they must integrate the lessons of psychology without reducing them to numbered brain areas. Psychologists in turn know that their subject has been transformed by the findings of neuroscience. If they are to help us to change our behaviour, cure our neuroses, think our way to better mental health, improve our lifestyle, enhance our self-understanding, or chase that elusive bird on the wing, happiness, they cannot ignore the lessons coming out of neuroscience.

Neuroscientists are aware of the need to ground their research in the shared values of the wider culture. Given the speed of advances in biotechnology, they have to take a clear stand as a profession on troubling moral conundrums that are no longer science fiction, such as selecting for intelligence in the womb, IQ enhancement for adults, prolonging brain life indefinitely, opening our private thoughts to all and sundry, and gifting human smartness to machines.

The discovery of neuroplasticity
There are many positive advances, such as the ability to save a life by excising a malignant brain tumour. Then there is the concept of neuroplasticity, or the idea that the brain has the potential to reshape itself. When our brain is young, many of our neurons are pleni-potential, or capable of taking on a number of roles. If we are born blind, our visual cortex can be assigned to other tasks.

This repurposing can happen because the same wiring patterns and codes are repeated throughout the brain. In 2000, the brain of a ferret was rewired, so that its eyes fed to its ear receptors, and vice versa, the animal soon learning to adjust to its new neural layout with no ill effects. Human neurons are not quite so flexible, but some stroke sufferers are able to make remarkable recoveries, reassigning old duties to new undamaged areas.

This both confirms and challenges the doctrine of localism, or the idea that brain operations take place in specific rooms. Some do, but the brain has to run a dual economy. Some areas contain 'slave' neurons that feed specific sensory input into regions of 'master' neurons, which are multimodal, capable of processing input from any sense. Walking calls on a co-ordination of skills

49

distributed across the whole brain, such as balance, skin pressure and awareness of where the body is in space.

The brain also has to run parallel processing, or performing several operations at once if we are to see, taste and feel the food in our mouth simultaneously. There is a lot of processing activity to cram into a small space. Computers running on linear algorithms lack this multi-tasking plasticity, making them liable to 'crash'. Our brain is more flexible: if one part 'goes down', others can continue to function. We can be blind but not deaf, and vice versa. Following a brain injury which damages our visual system, we can retain the ability to see an object in the distance, but lose the capacity to sense its movement.

The brain is therefore neither a blob of equipotential jelly with all areas quivering all the time, nor a machine with every part precisely labelled and waiting to be fired into action. It is a complex mix of localism and holism, functionalism and connectionism. It is both bound and free, structured and loose, active and quiescent. It hums along even when there is no stimulus, and its secret is its connectivity, all parts in touch with each other all of the time.

We see this most clearly in cases of brain damage. Surviving brain areas can deputise to a certain extent, allowing the brain a kind of self-repair that was previously thought impossible. Those born blind at birth can exploit their powerful redundant visual cortex to enhance their other senses, 'seeing' like bats through a kind of echo-location, or developing such a subtle sense of hearing that they can tune a piano over the telephone.

'My Left Foot' is the autobiography of Christy Brown. He was born with cerebral palsy which denied him the power of speech and the use of his hands, but he learned to use his toes as fingers so that he could write his story. In neurological terms, he wrote a new 'toe map' over his idle 'finger map'.

Allied to the exciting discovery of neuroplasticity, neuroscience has exploded the notion that we are born with a fixed number of neurons and incapable of growing new ones. In 1913 Cajal insisted that, after adolescence, the brain is limited for life to the quantity of neurons installed at birth, and unable to change the ones it has. It has long been known that salamanders can grow new limbs if necessary, but it was assumed that higher mammals like us with more complex neural systems cannot replace motor neurons, and certainly not brain cells.

It turns out that Cajal was mistaken. Some cells in our body are with us for life, such as our visual cortex, but others are replaced every few weeks, not just skin cells and bone cells, but also brain cells. Our hippocampus starts to shrink a little after the age of forty, but if we keep it up to the mark through exercise, it can make up to seven hundred new neurons a day, and even in our dotage as many as a hundred. By the time we are fifty, up to a third of our brain is younger than we are.

In a process called neurogenesis, astrocytes or star-shaped cells act as progenitors, prompting the growth of new brain cells. These allow dendritic branches to be regrown, synapses to be reconnected and circuitry to be rewired at any stage, a process so vital to brain health that conditions ranging from autism to psychopathy have been attributed to its malfunction.

This news is a boon for late learners, recovering addicts, depressives and therapists helping clients to break harmful behaviour patterns. It is not however a promise to return the brain to the elasticity, excitement and inventiveness of youth, or a reprieve from death. We cannot make the sun stand still, but we can, as the poet Andrew Marvell urged us, give it a run for its money.

As well as new cells being made, the ones we are born with continue to divide well into old age. Our hair and nails need regular cutting, and our wounds continue to heal, even if more slowly. Our red blood cells are replaced in our bone marrow every three months, the lining of our lungs every two weeks, our skin every month, our liver every year, and our skeleton every ten years.

The lining of our nose is an important area for growing new cells, possibly because our olfactory neurons, exposed to the wear and tear of the elements, need a constant capacity to self-renew. An unexpected bonus of our sense of smell might therefore be the stimulation of new brain growth, so every time we nose a rose or blink at a stink, we are also giving our brain a boost.

New neurons for old
Fasting can also give us a brain boost. Rats and mice kept on reduced rations grow new brain cells, as if their brain is thrown into survival mode, making new neurons as a response to tougher times ahead requiring smarter responses.

Even stroke victims demonstrate a potential for renewal. Depending on how many of their brain cells are damaged in the attack, and whether the control centre remains intact, they can 'rewire' undamaged parts of their brain to take over vital functions that have been lost before 'learned non-use' sets in. By the same token, trauma victims can 'rewrite' scarred memory tissue, because the economy of the brain is more about networks and dynamics than regions and mechanics.

In healthy brains, the implications of neuroplasticity and neurogenesis are enormous for self-reinvention, lifelong learning, reforming bad habits, correcting errors, creating a different future, changing minds, reviewing prejudices and starting to think positively, the potential benefits only just beginning to seep into education, therapy, policy making, public consciousness and criminal reform.

The leopard cannot change its spots, but it can choose where to sleep, or which prey to chase. No pattern of behaviour is fixed for life. The addict can self-transform, the sinner can find redemption, intelligence is not immutable, and there is hope after losing a loved one, our job or our home. Violent psychopaths who have been given life sentences have been known to leave prison and make valuable contributions to society.

Ignatius of Loyola, founder of the Jesuit movement six hundred years ago, famously boasted 'Give me a child until he is seven, and he is mine for life'. He is only partly right. While it is true that ninety per cent of our final brain architecture is set up by this age, there is much still to play for. This is because nature and nurture combine through and during the participatory role we play in the world: we act and are acted on.

Many come to religion late in life, while those who become disillusioned give it up. Those who perform martial arts become very aware of the power of their mind to shape their brain, while those who lose their physical mobility discover their ability to reset their brain's relation to their body, so long as they do their physiotherapy exercises. In a wider sense, every time we change our mind, we change our brain, and vice versa.

We possess this capacity for self-reinvention because our brain has a level of complexity not found anywhere else in nature. In biology, complexity is about more than being complicated. It's about generativity, not just the making of new lamps from bits of old lamps, but coming up with totally new ways of lighting up the dark that could not have been predicted from the few scraps we started out with. Our brain is not an input/output device like a computer, more like an orchestra, capable of making ever-changing music from the same notes, subtly altering melody, tone, tempo, key signature and instrumentation. One day it might suddenly come up with a burst of jazz.

Our brain is at once integrated and differentiated, ordered and disordered, stable and inconstant. Finding out how it pulls off this magic will dominate twenty-first century brain research, with huge prizes in store for the designers of artificial intelligence. No wonder as early as the eighteenth century, the Baron de Montesquieu called us 'that flexible being'. But there are limits: though Phineas's brain kept his body alive and allowed him to communicate normally, his frontal cortex was too traumatised for him to reclaim his personality fully.

How is our brain a hive of connectivity?

The neuron - the synapse – the neurotransmitter - the neurome - the connectome – emergence – the self

- A neuron is one of a hundred billion highly specialised brain cells found nowhere else in the body, which together make up our neurome.
- Each neuron is connected to its neighbours by tentacle-like structures called dendrites, and to the nervous system of the body by axons.
- The point of contact between each neuron is the synapse, a tiny gap across which complex molecules called neurotransmitters send electrochemical 'go' or 'stop' messages.
- By a process not yet understood, mind and consciousness 'emerge' from the operations of our brain.
- The moment we are born, we start to populate our neurome with our unique experience, to make up our connectome.
- The puzzle for neuroscientists is that they cannot explain how or where in this flurry of activity our sense of 'self' happens.
- This is partly because our sense of being a person in the midst of an engrossing life-world is not a fiction of our brain. It is a true story of our social and cultural existence.

The knot of nerves

Neuroscience demands a whole new language to explain itself. Neurons are cords that bind the sacred knot of nerves (a related word) at the top on the spinal column to the rest of our nervous system, operating as a super-highway.

There are over two hundred types of cell in our body, but neurons are highly adapted and finely specialised super-cells, the vast majority of them in our brain. They don't 'contain' thoughts or even generate them, but without their ministrations, our mind would be blank and thoughtless. They appeared about half a billion years ago in the evolutionary story, making big and fast brains like ours possible through their ability not only to store electrical signals but also to network and distribute them quickly around the system.

Each is a mini-computer, coded to perform a specific task by natural selection. It achieves its potential only alongside its billions of neighbours, as a tiny member of a much larger whole, a holon within a holarchy. The whole is contained in the part, but without the whole, the part is powerless. The whole of our genome is written in every cell in our body, but the whole of our mind is more than the sum total of all our neurons, however we add them up.

This does not stop each neuron from trying to maximise its influence in the brain, without making it selfish. Like an ant, it is programmed to work for the good of the colony. The mystery is that, although each neuron lacks comprehension, or doesn't 'know' who we are or what are up to, together they make up the phenomenon we know and experience as our waking self.

So far seventy five different types of brain cell have been identified in our neocortex alone. We are born with these, and if we are lucky we die with them, their axons or connecting threads stretching from our brain to our toes, through the half a million miles of fibres that make up our nervous system. They split into dendrites when they reach their destination, where they fire across synapses to control operations from talking to walking, gesticulating to urinating. There is also a 'second brain' of neurons in our gut that self-renew every two weeks, helping us to think as a whole body by coordinating our immune system.

The neurons in our brain are often stacked in columns, acting as pattern-matching switches and map-making transistors. There are eighty five billion of them in all, each a little brain in its own right, with no two alike. They are 'irritable', or set on a hair trigger, designed to respond to the slightest input, something we become very aware of during a horror movie.

Neuroscientists have established that certain neurons fire when we see an object turn left, others when we turn right, suggesting that making a decision 'up top' happens only when a critical mass is reached at the neural level 'down below'.

But they have not found what is jokingly called the 'Jennifer Anniston' neuron. It is tempting to think that everything we know about her is stored in a single cell, but the sight of her fires a range of neurons, because 'she' activates many brain areas: face and voice recognition, the sound of her name, memories of her on screen, the context we think of her in. It is a puzzle how our brain collates all this information, because our neurons are not neatly arranged in alphabetical order as in a reference library.

Each of our neurons does however have a 'job for life', containing a history of who we are and what we have learned. Brain neurons are not capable of replicating themselves by simple cell division, so it pays to keep our brain healthy, hanging on to as many brain cells as normal wear and tear allows.

Identifying from their billions how neurons carry out their specific roles or cooperate with each other is the principal challenge of cognitive neuroscience. Microscopy and photo-optics can shine a light on individual neurons, but the brain is so dense that it is difficult to peer deep inside to see how neurons form their tight networks. To give some idea of the scale of the challenge, each neuron expresses up to twenty thousand different genes, and no two are the same.

Although Freud did no experimental work of his own, he speculated that there are three key types of neuron: phi, processing sensation, psi, handling memory, and omega, modulating conscious awareness. Since then, many specialised cells have been identified in the brain, such as econoneurons that act as 'grid and place' cells. Many species of insect go on long foraging journeys, but manage to navigate their way back to base, and we too can find our way home even if we have been dropped in a different part of town.

Econoneurons perform many other functions too, such as speeding up our reaction to input, controlling response and enabling intuition. The brain's power derives from the fact that not only do neurons never work singly, they also often perform different roles or deputise for each other via vast interconnected networks.

A neuron is for life
Our brain is capable of making some new cells during adulthood in the hippocampus, mainly for memory purposes. The vast majority of our adult neurons were formed in the womb at a rate of millions an hour during our time of maximum brain development, highly dependent on our mother's low stress level and quality of nutrition during pregnancy.

Some neurons we lose naturally as we age, and some are 'pruned' by the brain as it streamlines its own operations in our first years. But the brain of an octogenarian can still be as bright as a button, assuming it has avoided the scourge of Alzheimer's disease, which randomly destroys vast swathes of irreplaceable neurons.

Each neuron is far more than an on/off switch, more like an excitable readiness for action in a state of superposition, uncertain of its next move until swept along in a microsecond by the energy of its thousands of neighbours. It can perform as many as a hundred thousand different brain functions, combining to give us the dazzling show we experience as our mental life. This fact alone should make us cautious about ever claiming to 'know' our brain.

As with our genome, 'mapping' our neurome still leaves us a long way from understanding how it operates. Some neurons, mainly concentrated in our left hemisphere, are devoted to specific tasks in the brain, while others in the right are more open choice. But whatever their roles or options, they form powerful alliances with their neighbours. They work, to borrow a phrase from Shakespeare, 'not in single spies, but in battalions'. Battalions fight in armies, which achieve high levels of coordinated and synchronised activity without it being obvious who is in charge, a feat emulated by the billions of neurons in our brain.

Neuroconnectivity

It's not just about numbers. We have as many cells in our liver as in our brain. What makes brain cells unique is their ability to talk to each other, with up to ten thousand links to their neighbours. Neuroconnectivity is as important as neuroplasticity, allowing knowledge and wisdom to be spread all over the brain. Experiments on rats show that removing parts of their brain after they have learned their way around a maze does not mean they get lost next time. They call upon memories stored in other brain areas, and other ways of knowing.

Connectivity, or the ability to distribute data right across the brain, is therefore the brain's secret weapon, explaining another great mystery: how the brain makes meaning. By what strange magic is an electrochemical signal transformed via a sensation into an experience?

A lone neuron cannot think for itself: it merely receives inputs, 'stupid' in the sense that they could be about absolutely anything. The Holy Grail for researchers into artificial intelligence is how these are converted into intelligent outputs. How does our brain form ever denser networks, levels of organisation and hierarchies of command with no obvious central commander?

Together our neurons form a complex adaptive system, like a giant termite mound, their unpredictable connectivity faster than all the world's financial markets, and just as turbulent. In their billions, constantly firing and dying away, they generate an emergent property much greater than the sum of its parts, like a cloud from water droplets. Without such a phenomenon we could have no sense of a self, or the irreversible flow of the arrow of time, or the piquancy of the moment.

Post-Darwin, we have come to see the brain as the product of millions of years of natural selection, not divine design. The paradox is that, despite the absence of an intelligent creator, the *result* is a highly intelligent brain. Out of chaos comes cohesion, out of simplicity comes complexity, out of stupidity comes smartness.

The key point to remember from a Darwinian perspective is that, as better brains were selected in an evolutionary ratchet principle, the end products of a clever brain and the wonder of consciousness could not be foreseen at the outset. Instead, each neuron is there because it allows its owner to live another day, as part of a cooperative of sure-fire neural programs selected over thousands of generations. This allows us not only to smell the rose, but also to write sonnets about its beauty.

Each neuron is there because it *works*, or is 'good enough' to do its job, and even if it misses a trick, it has thousands of back-ups. Also, its effectiveness is increased exponentially when it cooperates with its neighbours. We might say that 'selfish' neurons use human brains as carriers to make more of themselves, but no matter so long as more self-aware humans survive to the next generation.

This 'neuron-first' calculus means there is no controlling 'centre' in the brain, or master genes, or glowing hot-spot where God or the self resides. Instead, we owe our sense of possessing a whole brain to the neurons connecting up their inputs and outputs all of the time, ceaselessly talking to each other as messages fire to and fro.

Signals coming into the brain are a one-way street, but *within* the brain, communication is two-way, without which we could not formulate return messages. An added complication is that the brain has multiple pathways: if a message is blocked along one route, there are thousands of others in readiness, making it impossible to predict any particular brain state or response. Each neuron listens out for news of a change, like a spider sensing all the strands of its web. But there is no spider, only coherence and unity which is greater than the sum of the parts. The brain always goes beyond what is given to it.

Neural handshakes

Quite how our brain achieves this feat requires a bit of imagination. Think of our neurome as a football crowd all holding hands (unlikely I agree, but go with the analogy). The fans are individual neurons, free but carried along by the crowd, housing their own memory and knowhow, microcosms in a macrocosm, junctions in a giant web. All the arms reaching out are dendrites, from Greek for tree, spreading like branches, eager for contact. Even in a huge football crowd we, like our neurons, enjoy only 'six degrees of separation', putting us potentially in touch with everyone else in the stadium.

Each touching hand in the crowd represents a neural handshake or synapse, from the Greek for joint. This gap is only a ten-thousandth of a millimetre wide, across which 'transporter' neurotransmitters ferry electrochemical messages back and forth, plugging everyone into reciprocal circuits. Ions or charged atoms spark 'action potential', causing a 'spike' of information exchange through sodium and potassium 'gates' at up to three hundred feet per second.

It's not just about speed: connectivity is the key. Each neuron can 'talk' to others in the brain through up to ten thousand synaptic links, giving a total of a thousand trillion potential connections across the whole brain. 'Speed of thought' is really therefore a matter of tight wiring. Some synapses are further away from each other than others, but what matters is close teamwork between neural webs. Some researchers put high IQ down to speed and proximity of synapses, but they forget that we also have other 'intelligences' that rely on more leisurely reflection, which is why the answer to a problem often comes to us in our sleep or daydreams.

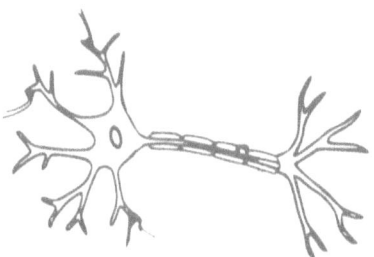

The spidery neuron
This sketch of a neuron gives some idea of the billions of connections it is capable of making with its millions of neighbours in a giant neural web.

We Are More Than Our Brains

In some ways the tiny gap between our synapses slows down our processing speed. The brain runs not just on 'dry' electricity but also 'wet' biology, because it is a living organism, not a computer. It can't crunch data with the rapidity and efficiency of our iphone, but by sacrificing a little speed, it gains subtlety and diversity of response, which are the foundations of our freedom and creativity.

Neuronal mechanics are at work in the sense of forces, resistances and motions, but chemistry has to weave its spells too, breaking down the catalysts in the neurotransmitters to send them their different ways, to achieve their different effects all over the brain.

In other words, our tiny synaptic gap is what makes us human. It is in the millisecond between stimulus and response where mind and individuality occur, the basis of our claim to any power of free will.

Our synapses possess action potential, either as exciters (go ahead) or inhibitors (hold back), fired or not fired by up to fifty different types of special proteins released by neurotransmitters and neuromodulators which speed up, slow or block the flow of traffic across the synapse. The most important of these, serotonin, dopamine and norepinephrine, collectively called monoamines, were discovered and named in the 1950's, and their subtle biochemistry is still being unravelled.

Potentiating synapses
Neuroscientists have discovered that our synaptic traffic is moderated by a push-me-pull-you mechanism of long-term potentiation (LTP) and long-term depression (LTD). Those with a good memory possess strong synaptic potentiation. At the opposite extreme are those with depressed synaptic activity who lose whole aspects of their past, or are unable to lay down new memories.

An imbalance of synaptic potentiation and depression, with some gates in the brain slammed shut or stuck open, may lie at the root of obsessive compulsive disorder, Tourette's syndrome and chronic pain. It may explain why recovery from trauma is so difficult: bad memories are triggered too readily, and better ones are hard to superimpose. It accounts for aspects of addiction: repeated actions, intense experiences and dopamine-inducing rewards increase the likelihood that synapses will fire too quickly and easily next time, made worse by a weakening of the ability to take the finger off the trigger.

Depressed synaptic activity therefore makes it hard to inhibit unwanted response to stimuli, or break ingrained cycles of thought and behaviour. On the positive side however, synaptic gates that are highly potentiated, left open to novelty and change, are a boon for learning and creativity, especially in our early years, and even in our eighties.

The vast majority of our brain activity is non-conscious, designed to keep us alive, regulated by hormones that are slowly released by our body clock, or fast-fed by a threatening bang behind us. This does not mean that our mental life is 'automatic' in any sense. It is in our healthy conscious moments that we become aware that our synapses are not just slavish yes/no mechanisms, but active agents determining our moods and insights.

When we feel moody or angry, we become more aware of the sudden surge or jam of neural traffic in our brain. It is at such times that we can walk away, change the subject or breathe deeply to consider a more rational response, which is why some commentators believe that understanding the synapse holds the key to human nature.

Every time we make a decision, the nine possible responses that are depressed, blocked or inhibited in our synapses are as important as the one that potentiates us, or that we choose to give expression to. When we see a goal scored, our neurotransmission will greatly depend on whether our team has just won the cup, lost it, or we don't care either way.

The society of neurons

So, back to our football crowd. Now we have to ramp up our imagination massively in scale, because this huge gathering of fans is only one tiny neural net in the brain, confined to one stadium in two dimensions. It is wired up in three dimensions to millions of football crowds supporting different teams all over the globe. Football fans of the world, unite.

If that sounds big, it's still not big enough: now we have to think *really* big. Towards the end of its development in the womb, the brain of a foetus will be adding a quarter of a million brain cells *every minute*, and at birth will have a full complement of nearly a hundred billion, all waiting to be wired up to each other the moment it hears its first sensory input.

Nine tenths of these cells will be glial cells, from Greek for glue, providing vital services to the remaining billions. Unlike the 'main act' neurons they are not electrically active, but neuroscientists are steadily establishing how vital their supportive role is in removing toxins, assisting the neurotransmitters, and directing the pruning process as the brain grows.

But it doesn't even stop there. Each football crowd is unique, supporting a different team, attracting a particular fan-base, swearing allegiance for personal reasons, enjoying a peculiar record of victories and losses, and these experiences unite biology and culture, personalising the neurome into a connectome to the point where nature and nurture meld into each other.

The mind-crowd remains elusive, because when the big game is over, the spectators disperse, going on countless journeys home to particular families by different means, each fan becoming an individual again. The crowd exists with a mind of its own only while the match lasts, after which it goes its separate ways, just so many scattered ticket-holders with private allegiances. It's as if each thought we have is a temporary gathering which vanishes as soon as it assembles.

This contrast between the parts and the whole, the momentary and the permanent, is a challenge for neuroscience. It's one thing to research individual neurons minutely for years, another to understand how each cell sacrifices its autonomy to the greater cause, joining the flow of the whole body. The neuron observed by the neuroscientist, even as part of a wider network, can never amount to the message it carries in the life of a busy mind.

This is why the mind is so hard to locate or explain: the brain is a society of neurons, each with a mind of its own, but allied to the wider cause, never totally smart but capable of subtle response. This makes the brain a grand confederacy and abstraction, coming together only when occasion demands.

We don't find society by looking for it on a crowded street: we are already part of it. The same is true of mind: it is neither located in our neurons nor a disembodied essence. It is our every thought for as long as it lasts. We cannot see reality any other way.

The miracle of emergence
Some neuroscientists are so wedded to chains of cause and effect that they are stumped to explain how the phenomena of mind, self and consciousness come about, except as serendipitous outcomes of information constantly streaming through endless neural circuits.

The challenge is that, in accounting for the presence of mind in the world, we find ourselves appealing not to the laws of physics, but to the vagaries of biology, where mechanical models do not rule the roost. In the macroworld of rocks and trees, we can predict outcomes: dropped stones and leaves *always* fall to earth. The microworld of brain activity does not allow such certainty. Billions of moving parts, bouncing this way and that off each other like demented billiard balls, add infinite layers of unpredictability.

When we scale up this level of uncertainty to the human world, it feels like we have arrived at the edge of chaos: *sometimes* she likes the presents we give her, and *perhaps* he will forgive us for offending him, but we have no way of being sure.

Given such complexity and unpredictability, some neuroscientists appeal to the idea of emergence, or supervenience, to account for how we get from millions of stupid inputs to one overpowering intelligent output. How does the brain organise matter into mind, convert mechanism into meaning, and turn competence into comprehension?

The theory of emergence suggests that the brain's high level of connectedness evolved exponentially over billions of years. Its premise is that the urge to complexity is intrinsic to life itself. When a few cells are placed in a Petri dish, they either wriggle around looking for food, or band together to form a primitive life form such as a slime mould. As the levels of complexity escalate, it is almost inevitable that a kind of mind will emerge.

Once nervous systems developed, brain cells proved to be even more other-oriented. Placed close to each other in the laboratory, they begin to oscillate in unison with their neighbours. In this way a proto-brain is formed. The next step is for each cell to take on a specialised role in the service of the whole. Gradually intelligence, consciousness and agency begin to 'emerge', just as a crystal 'appears' from cooling liquids, a storm from roiling molecules in the air, language from symbolic grunts, culture from shared acts of learning, and law from public contracts. Whole systems emerge from parts, unfolding into the wider world.

The key point about emergence is that, like evolution, it doesn't need a driver, or an end goal, because single parts going about their work locally and individually can make global networks and integrated systems. A mind-brain is merely the serendipitous outcome of any bunch of cells packed together with high information content, not necessarily inevitable or a human birth-right.

This dents the pride of those who see human minding as somehow divinely ordained, specially added or superior to anything else in nature. But there is no need to appeal to a Creator or Prime Mover: biodata slowly morphs into meaning, on an ascending scale, until we end up with intelligences like ours debating mysteries like this. There is no magic moment when mind appears, just a tipping point before which there is no mentality, and after which there is mind. One day the universe is not aware of itself, the next it has created its own consciousness-appreciation society, which is not to say that it knew we were coming.

The emergence of mind is a miracle nonetheless, albeit a natural one. On their own, neurons are blind and dumb, but when they combine, they make an entelechy, or a new kind of intelligence, with a potential that is infinitely greater than the sum of the parts. They work as a giant collective, spreading axons exponentially in every direction, making networks so dense that they can process their operations not just serially but in parallel, which computers currently struggle to do without crashing.

They become the chief executives of a superorganism, allowing us to multi-task: we can stand up, listen to a piece of music, hold a conversation, peel the potatoes and maintain our body temperature, all at the same time. Not only that, our eighty five billion neurons convince us that the world fits itself to our mind, not that our mind evolved to fit itself to the world. From our conscious perspective, we are mind-centric, and reality is as we see it.

Society of mind

The 'emergence' of our mind does not exist independently of its component bits, but neither is it reducible to them, or predictable from them. As we ascend the hierarchy of explanations, from single neuron to society of mind, we can see that each step hints at the next, but we can't reverse this logic. A bird flying overhead is not reducible to a single 'cause' of its being aloft, but a complex of many phenomena: aerodynamic properties, bone structure, feeding habits, reptilian ancestry, mating behaviour, the quantum mechanics that keep it from disintegrating.

Each of these is a feature of the whole, working perfectly to its own evolutionary logic, but cooperating at every level, from the tiniest particle up to the bird's silhouette against the sky. None of the separate parts, or even the fully assembled article, 'explains' what a bird is. As Dorothy Parker might have put it, a bird is a bird is a bird is a bird.

Just as billions of hydrogen and oxygen atoms combine to give us the beauty of water, so trillions of synapses super-convene to give us a new form in nature, the society of mind, different not just in degree, but also in kind from the ingredients that went into its making.

Another way of looking at this is to say the mind is to the brain as the flower is to the plant. Nature did not set out to create a thing of beauty or fragrance, but colour, scent and petal shape, intent on attracting pollinators to make seed for the next generation, combine to give the gardener an unlooked for delight.

Given the phenomenon of emergence, there is nothing to be gained from muddling levels of explanation, or claiming that any higher form can be reduced to its zillions of parts. This is why 'substance dualism' appeals to many neuroscientists.

This allows an acceptance that mind is not just the brain, while insisting that it is highly dependent upon it. The brain doesn't cause the mind, but it facilitates levels of interdependent organisation without which the mind is a vehicle with no passengers.

Traffic cannot exist without cars, but until the cars embark on their separate journeys, we cannot predict how the traffic will behave. There might be a totally unexpected accident, perhaps involving us. We could get caught in a queue, even a jam, or we might get to our destination without a hitch. Our mind is a similar daily pilgrimage, on which we don't know how the journey will go until we set out.

Network theory

Network theory is an attempt to account for the emergence of mind through 'bottom up' causation. Just like the waves on the surface of the sea, the brain creates pattern or synchrony out of all the roiling activity below. A surfing wave is a naturally occurring 'hub' or concentration of energy which makes the ocean *look* regular and organised on top. We then attribute 'top down' causation to these waves, as if they are controlling what goes on down in the cradle of the deep: they rock ships, erode cliffs, crash against reefs, carry surfers to shore, and leave patterned ridges of sand on the beach.

But they're only waves, or bundles of energy, or surface signs of something vastly more powerful going on out of sight. This leads some researchers to see the mind and the self as floating froth, the random effusion of thousands of toiling networks, not 'real' except as a passing illusion or unlooked-for emergence, like the genie appearing from Aladdin's lamp.

The mind and the self might be similarly serendipitous. To our consciousness, their 'essence' feels real enough, but they might be no more than happy accidents of something designed for a totally different purpose. In a building, there are many features that are incidental to the main structure. A wall serves the primary purpose of supporting the roof, but it also gives us an opportunity to show off some lovely wallpaper or photographs. Nevertheless the real load-bearer is the brickwork, not the picture gallery that hangs from it. In this sense, the mind and self can be seen as little more than the brain's decoration.

We experience the phenomenon of emergence in other areas of our lives. A city emerges from a maze of streets, the economy from thousands of purchases, weather from the chaos of wind movement, the internet from millions of posts and tweets, the surface tension of water from the behaviour of molecules out of sight down below.

There is organisation without an organiser, a new property that is dependent on its cause, but not reducible to it. Instead of uncoordinated activity, there is the super-organism of a body in touch with all its parts, its arms and legs not flailing independently but harmonising for the greater good.

Another aspect of emergence is that it is uniquely human. Without getting into deep water over the nature of 'reality', we have to remember that our brain is the continuity editor of our skull cinema, blending the pixels and splicing the rushes into one, long, seamless 'take'. It evolved as a causal dynamic modeller. What appears random must be made sense of, and its origin accounted for. Out of all the bitty inputs it receives, its instinct is to paint a whole picture, filling in the gaps, connecting all the networks.

As a result, we can't see reality any other way, except as a fait accompli of cause and effect given to us by our brain. The way things appear to us is how they are, and vice versa, and we can't get behind this shadow-play. Emergence might create organisation without an organiser, but for there to be meaning, there must be a meaning-maker.

From neurome to connectome

Wherever and however they emerge, or are accounted for, the mind, self and consciousness, like thought itself, do not happen in specific areas of the brain. There may be 'hot spots' that show up on a scan, but our sense of being is distributed through every brain cell.

Consider for instance what happens when we think of 'the dog that lives next door'. There is no single neuron that stores this idea, or remembers the dog's name. Taking a dog for a walk calls on tens of thousands of neurons talking to each other in vast interconnected networks and abstract patterns of hierarchical command-and-control centres called ganglions.

These tightly knit communities of neurons process information both one-at-a-time serially and several-at-once in parallel, linking perception, memory and thought in infinitely novel ways. Why has

66

the dog next door been barking so much this morning? Is it a bark of warning or distress? Should we go round and check? More unusually, if no-one is at home, why has the dog suddenly *stopped* barking?

This diversity of operation, flexibility of response and power of combination is what allows our brain to give us a totally convincing but highly individual experience of reality. In their billions, our neurons make up our neurome, just as our genes make up our genome. But neurons, just like genes, have to be expressed in a particular life-world, giving each of us a connectome or unique sense of being in the world.

It doesn't really matter what name we give to our sense of being here now, or what account we give of how our brain generates it, despite the fact that neurophilosophers debate these issues long and hard. So long as we remember that our connectome is a hierarchy of operations, from the tiniest neuron to our whole nervous system interacting with those around us, and that our individual personality, the quality of our mind, our sense of agency, the tang of our dreams and the ache of our desires arise from its hyper-networks, we will avoid making a category error, or talking about ourselves at the wrong level of explanation.

The self under attack

The idea of the mind and self as tricks of the brain is unsettling to a generation brought up in a Judaeo-Christian tradition of a free-choosing self, fed on a diet of individualism and self-realisation, but the notion of the humanist, Romantic, postmodern or digital self as different faces we show to the world is of no use to neuroscientists, nor is fuzzy talk about inner thoughts, mental states, identity crises and metaphysical souls.

The crisp brain-speak of hard-wiring, neural networks and synaptic clean-ups offers them much greater clarity than woolly thoughts, malleable moods and psychobabble. Empirical science bent on causal reality has no time for the kind of fanciful introspection found in novels or diaries, where the mind goes on a ramble, and the self comes across as a fickle confection projected by our consciousness to get us through the day.

Neuroscientists point to the phenomenon of depersonalisation, or the feeling of being outside of ourselves, as evidence of the fragility of the self. We experience this psychic disintegration in the seconds before an accident, or when we think we are going to die. The rickety frame that holds our sense of self together deep inside the insula temporarily falls apart. For a few the sense of dislocation and detachment from their own

67

feelings can be chronic, leaving them with a sense of numbness and lack of identity.

Evolutionary psychologists question the idea of the sovereign self from the perspective of deep time. Our brain has evolved as a smart assemblage of ad-hoc solutions, not a paragon of design. As a result, our mind is not essentially wise, our nature is not naturally virtuous, and our sense of self is not philosophically profound. Social psychologists agree. It is 'flat', susceptible to the nudge and the hidden urge, denied the depth that we used to call character, psyche or spirit.

Ironically, support for the debunking of human specialness has also come from the arts and humanities. In the 1980's postmodern theorists began to deconstruct the foundations on which we build our notions of liberty, progress, reason and humanity. It was only a short step to apply their sceptical approach to the mind, self, culture and spirituality.

Most concluded that the mind is a flickering screen and the self is a shallow illusion, but for very different reasons from neuroscientists. They insist we are not free agents capable of 'uncaused' creativity and self expression. We are trapped in our prejudices, prisoners of our language, slaves to our belief systems, figments of adopted personae, commodities of our culture, ciphers of our digital codes, projections of the signs we give off, products of what we consume.

There is no deep identity, essential 'me' or stable 'inside', only a fluctuating outside that demands a regular update if not total reinvention. To this end, we can be what we dress up as, retro steampunk, futuristic cyberpunk or almost-now dieselpunk.

And yet art, literature and drama are about nothing if not exploration of the 'lived' self, real or imagined. If the self is a chimera, how do we account for the fascination of Shakespeare's great tragic heroes, tormented as they are by jealousy (Othello), revenge (Hamlet), lack of self knowledge (Lear) and ambition (Macbeth)? When we see these plays in the theatre, we are overwhelmed by the sense of a once-strong self slowly unravelling.

These are fictional characters, but their suffering touches a deep spring inside us, as does music. The composers of the great symphonies are real people, expressing intense feeling, but their works are empty if the sounds resonating in our ears have no emotional self to relate to.

We can enjoy art of all kinds because we possess empathy, or the ability to enter the experience of another mind, albeit temporarily. Actors are skilled not only at adopting another identity, but luring us into it. Scans show that when we enter a role in this way, or read a novel, or inhabit an out-of-body avatar in a computer game, the 'self' area of our brain quietens down, as if briefly relocated.

This revelation of neuroscience does not undermine the idea of self. It underlines the strength of the self we start from, and return to as a changed person. After our multiphrenic explorations of other people's mind states, we revert to the idea of our own self as a primary act of mind, albeit transformed. Without it, we are zombies, unable to override our impulses, alter the course of our lives or impose our will on nature.

The riddle of the self

Philosophers in the Western tradition have nevertheless struggled to define the nature of the self, and whether it exists at all. Descartes claimed 'I think, therefore I am', convinced that when he introspected, he saw a stable and sovereign self. Around the same time however others were beginning to doubt the existence and continuity of a 'foundational' self.

In the seventeenth century, Sir Thomas Browne remarked that 'We are ghosts unto our own eyes'. When we introspect, we don't see a single, substantial, stable self, but a stream of intrusive unrelated thoughts. Perhaps therefore Descartes, for all his certainty about his selfhood, was just a collection of atoms that came together for a short while and then dissipated. After all, the young Descartes who served as a soldier had changed just about every cell in his body by the time he turned to philosophy. So was he the same person?

Psychology fares no better at explaining the self. Where now is the bundle of perceptions that was Descartes? To complicate matters further, he probably harboured a shadow self, the part of him that he repressed, ignored or never knew he had in the first place. The ultimate confusion is to speculate what would have happened to Descartes' self if his brain had been transferred to the body of a man called Francois. What name would he answer to?

Pragmatics comes to our rescue: there is no known society in history that has shown it can operate without a developed sense of self or shared concept of mind. Each is implicated in the other. We have evolved to be social creatures, not solitary thinkers, so it is to others that we owe the shape of our mind and our sense of self, however culturally moulded.

We Are More Than Our Brains

Sir Thomas Browne
1605-1682
'We are ghosts unto our own eyes'. When we look inside ourselves, what do we see? A stable self, or fleeting impressions?

Also, while we might get annoyed by our erratic flow of thoughts, we *need* our tiny sliver of consciousness to capture the *this*-ness and *now*-ness of each passing moment, of what it feels like to be *me*, alive and doing this activity. They may be fragile and fractured, but mind and consciousness are all we have to help us savour the past, feel the present and anticipate the future.

Materialists of whatever stripe cannot subvert our consciousness, if only because our socio-cultural mind is as 'evolved' as our biological body, going through the same tough selection process. We are *both* neural networks *and* creatures bristling with irrational beliefs, *both* a public biological brain *and* a private mind seeking meaning. Each of our thoughts may well have a neurochemical cause, but there are no specific neurons that stoke our awareness, or identifiable genes that encode our selfhood.

In any case, our genes and neurons never work singly but in complex combinations and subtle correlations with what we think and feel. Our consciousness is no more translatable to neurochemistry than a beautiful melody is reducible to sonic vibration.

This does not mean that our thoughts are not malleable. We are the thoughts we think, which is why psychotherapists insist that, rather than allowing our brain circuits to do our thinking for us, we can reprogram our thoughts to change the weather inside our brain.

Defending the self

Challenging our ego and sense of human uniqueness is good for us, whichever quarter it comes from, because it obliges us to examine our assumptions and question whether self-knowledge is possible. If the self is an illusion, why did it evolve? What are the consequences of deconstructing it without something equally powerful and cohesive to put in its place? Can neuroscience and evolutionary psychology give us a new understanding of the self?

Before we answer these questions, we need to remind ourselves that traditional 'belief in' the self has some robust foundations. Science is our key to understanding the physical world, but there are no formulae to explain our double historicity. We are born Homo sapiens, but to *become* a human *being*, we have to grow in two dimensions: as individuals on a unique life journey, and as members of a society saturated by politics and culture. Only in these realms can we express intentionality and agency. These are hallmarks of the self, neither of them as predictable as the movement of the stars.

Even if the self could be objectified, it is not single but multiple, because we inhabit a manifold intersubjective reality. We each carry multitudes and contradictions within our apparent singularity, by turns lovable rogue, gentle giant, soft disciplinarian, noble warrior or big baby. Sometimes, beneath our brash exterior, we can be all these things at once.

Also, although introspection has long been dismissed as too subjective for empirical study, we find that subjectivity can be objectified to a degree. We experience the self as a quality in our life, but we can also quantify it. When we read diaries and social media, or keep a record of our self-observed thoughts as part of a controlled study, we find that we monitor our moods, analyse our perceptions, track our ideas, and fine-tune our well-being in response to our bodily feedback. We discover a pattern in our minding, the same concerns and insights occurring, suggesting that there is a cross-cultural lived reality to the self that is consistent and predictable.

This data does not amount to a 'law of the self', but it does give grounds to believe that the self has a recognisable and replicable psychological form and function. Questioning its existence is tantamount to doubting the earth we stand on.

We know that we change as we grow, constructing new selves to greet the faces we meet, otherwise we cannot be navigators of social space, free choosers, captains of our soul, or bearers of rights and responsibilities. We face the eternal paradox of becoming a self while also learning to be selfless, of reconciling being someone who thinks with someone who feels. The whole point of morality is that, while we are creatures of nature, we can also detach ourselves from nature, choosing with others in mind, not just in our own interest.

Most researchers into the nature of the self realise this, but they often lose sight of the whole by seeing the self through too narrow a lens. Developmental psychologists divide our cognitive self into different kinds of intelligence, such as emotional, moral, spiritual and kinaesthetic. These might have evolved as separate mind-tools, but they don't occupy separate cubicles in our cortex. Our brain unifies them as joined-up experience, inferring connections and filling in gaps where necessary.

If that is a conjuring trick, it is the best kind of magic, because its whole point is to hide from us exactly how it is done. The self is indivisible, because it evolved to give us a seamless and integrated sense of being in the world.

The philosopher David Hume remarked that, when he looked inside himself, his awareness of 'self' was always linked to a sensory perception, or what his mind was paying attention to at that moment. There was no 'self' to look at, only fleeting impressions.

While he was looking however, there were other essential elements of his self going on quietly in the background: his memory, mood, imagination, sense of the moment, ability to reflect on the past and the future, and alertness to the presence of other minds.

The social self

Hume needed also to consider the self from an interpersonal perspective, as both a private and a public necessity. We need a still centre where we can anchor our point of view and our dealings with the world. We require psychological continuity in the form of the memories we carry from one day to the other. Even when we are stripped of our clothes and make-up, we still stand in our nakedness as a coherent person who perceives, thinks, believes and suffers, if only because we feel cold and embarrassed.

From the inside, the self is a private, existential reality, a verb and not a noun. It's something we *live*, in the first person. But it did not evolve this way. The self is primarily social, acted on from the outside. The child psychologist D W Winnicott pointed out there is no such thing as a baby. There is only a growing mind that is socially constructed by exchanges with parents, carers and other children. The self is a reflection of those we mix with, and how we have evolved to respond to the demands they make on us.

Society makes its mark on us in other ways. On one side is a workplace that encourages conformity and suppresses individuality, and a government which wants to know more about our self than we do. On the other side is a consumer culture that urges us to self-improve, self-realise and self-actualise, begging the question whether there is a core 'self' to work on in the first place.

Meanwhile our personality and abilities are put to the yardstick of psychometric testers and employers, who apply a 'technology of the self'. They want to type-cast us as thinkers or feelers, leaders or followers, extraverts or introverts, on the assumption that these traits pigeon-hole us in all situations, and are fixed for life.

It they are right, we are left with no scope for character, which is a measure of our freely-expressed virtues. There can be no epiphanies, revelations or fresh starts, because we have no means of turning vectors into values, memes into meanings or opportunities into transformations. There must therefore be more subtle aspects that we have yet to consider, of how the biology of our brain gives us the being of our mind.

How do we balance fixity with flexibility?

Neurogenesis - networking - synaesthesia - neuroplasticity - neural pruning

- We have the largest brain of any primate in relation to our body.
- We have to be born early with a soft skull to allow our brain to inflate to full size in our first year.
- There then follows an extended period of childhood learning, made necessary by the complexity of human culture.
- We are born with all the neurons we will require as an adult.
- To balance fixity and flexibility, some neurons need to be task-specific, while others must remain open to new learning.
- Some of us experience cross-sensory synaesthesia while our neurons settle down.
- In a counterintuitive process called neural pruning, on the principle that less is more, our brain maximises its learning by discarding underused neurons to make way for the neurons it wants to keep.
- At first our brain relies heavily on pre-programming, but as we learn, we begin to put a personal stamp on the kind of brain we will have as an adult.
- Adolescence is a time of excitement, uncertainty and risk while the teenage brain undergoes a major remodelling in preparation for adulthood.
- The degree of flexibility and openness to new experience that we retain as an adult is largely a reflection of our mindset and curiosity as we make the journey.

We Are More Than Our Brains

In the womb

The brain is very active two thirds through its time in the womb, making a hundred billion new connections each day, forming vast networks, especially in the last two months of gestation. Babies can't grow their brain to full size inside the womb, otherwise their heads would be too big to be born. Unable to pass safely through the birth canal, they would kill themselves and their mothers in the process. Human childbirth is therefore harder than for other primates, an evolutionary compromise between the baby's big brain and the mother's narrow pelvis. If her hips were any wider, her ease of locomotion would be compromised.

There must have been a time in our ancestry when women could give birth unaided, but no longer. The expertise and comfort of midwives are now seen as essential, as well as the occasional caesarean section, but this is a classic example of the law of unintended consequences. These supportive practices make birth easier but also harder for women in the long run, because they outwit the rigours of natural selection. A woman who struggled to give birth on the savannah would not have survived, and nor would her child.

All these factors put together mean that human babies are born early and vulnerable, requiring extra time for additional brain growth outside the womb. This explains why we are born with a soft skull of overlapping bones that can 'inflate' on entry into the world. At birth the spongy top of our head barely protects the delicate brain beneath, but gradually the bones of our cranium expand, fuse and harden.

Head-binding during the soft-boned early months seems to have been a common practice in Ancient Egypt, giving some mummified Pharaohs unusually elongated skull shapes, though their brain seems to have developed normally. Premature babies with delicate crania require extra special care to shield their vulnerable brain in the weeks after birth.

Assuming normal birth, we are born with all the neurons we are ever likely to have. But huge numbers of neurons alone do not make a brain. Three vital processes must occur for this to happen: the neurons must be allocated their roles, they must be connected up to each other in dense networks, and there must be input to trigger contact between them. As well as a medium, there must be a message, and the more the better. Even inside the womb the neurons are primed for stimulus, responding to sound and pressure. Locked away in the dark recesses of the skull, they are useless until they are activated by the light of experience.

We Are More Than Our Brains

The brain is born for culture as a fish is born for the water. By the end of this immersive process, if the links between our neurons could be shown as a map of the ever-flowing ocean currents around the globe, we would see every molecule of water bound to each other in three dimensions, with no inlet or rock pool left incommunicado.

Special delivery

Before each molecule is swept along, it needs to know its destination. At first in the womb, at the blastocyst stage, our cells are pluripotent, or capable of taking on any role. To build a body, multi-purpose stem cells must become differentiated tissue, which means assigning them specific jobs in the brain, heart or lungs. Each cell in our body therefore, and particularly in our brain, is a special delivery.

In ways scientists still do not understand, from the moment of the first division of the fertilised egg, cells seem able to arrive at the right address in the brain, and just as remarkably know how to participate in the making of a mind. This route-finding is directed by shape-generating genes, or morphogens, following instructions coded by up to a hundred thousand proteins. Just like salmon following a chemical trail from the open sea to the stream they were born in, so the cells in our body 'read' a trail of complex sugars that encode their final destiny.

Consider the precise relabelling that must go on in the internal gloop of a caterpillar to guide its metamorphosis into the complex imago of a butterfly. It is as if each cell already knows its place and function in the finished organism, a process which has been likened to a baby setting out on its own on all fours from Land's End and successfully finding its way to John O'Groats.

To help our cells get to the right address, especially our brain neurons, they are chemically tagged, unfolding to a strict program and schedule. There is no point building a body without a brain to control it, and vice versa. This explains how neurobiologists are able to advise law-makers that pain-sensitive nerve cells are not formed in the womb until after twenty four weeks, with the result that in some countries abortion is allowed up to that point but not beyond it.

The time-scale of the postal service of cell delivery is crucial for a full understanding of child development. Psychologists and teachers are now wary of labelling children as 'slow developers', though parents naturally worry if they feel that their child is 'behind'. So long as there are no underlying problems, the genes know in which order to construct the building, and what it should

76

look like when finished. The scaffolding will fall away naturally when it is no longer needed.

Parents notice growing spurts, developmental leaps and phase transitions in their babies, usually arriving on cue, sometimes early, sometimes delayed. No toddler will run before it has learned to walk. Our grandmothers were content to let their children grow at their own speed, but modern parents are made anxious by conflicting advice from how-to-raise-your-baby books. Should they feed on demand or to a schedule, liquids or solids, breast or bottle, swaddled or free, sleeping alone or with the parents, on the back or tummy, with white noise or in silence, in light or in darkness, controlled crying or rocking to sleep? Given this level of fretting and confusion, it seems a miracle that any of our ancestors made it through to adulthood.

Baby brains are difficult to put in a scanner, but neuroscientists have encouraged a wider acceptance of neurodiversity and rate of development, justified by the discovery that there is no single template for the construction of a brain. So long as the neurons arrive on time and in the right place, the brain finds unique ways of wiring them into networks based on experience. From the brain's perspective, and to the advantage of human communities, diversity is not only to be expected but also desirable.

As many know to their cost, there is a price to pay when the fine-tuning mechanisms of cell delivery falter, or when the proteins that control the process go rogue. Cells undergo mutations all the time, causing malfunctions which might drive evolution to the next level. Cells that cannot perform their proper role in the commonwealth of the body, dividing randomly and suffocating other cells, are the principal causes of cancer.

Staying flexible

Once in place, with few exceptions, the role our body and brain cells play is fixed for life. This is why, in the case of skin and limb transplants, the body might reject them. The brain won't recognise them until it reprograms them and rewrites its neural map to accommodate them.

In the brain however, as we have seen, there is an important capacity for fluidity amid the fixity, or what is called neuroplasticity. In a new and healthy brain, once cells have been assigned there, they have to specialise further, and to do this, they require more sophisticated protein-coding abilities than other cells in our body.

To achieve this, our neurons are stewed in a soup of a hundred thousand proteins and forty million amino acids. This gives our brain a hugely flexible mapping capacity, essential because evolution had to fix some specialist skills, but also enable flipping between a set of general rules and particular sensations.

Our brain cells are therefore more flexible than we think, not fixed like a computer program but capable of self-adjustment through feedback. This has led to a dream of medical science: to recapture the flexibility of young stem or 'master' cells whose DNA is not yet 'finalised', but capable of being programmed to repair damaged tissues in diseased brains, or make good the ravages of cancerous tumours. Alzheimer patients might be able to grow new memory cells so they do not forget the faces of their loved ones.

Cochlear implants already allow the brain to hear by bypassing damaged auditory nerves in the peripheral nervous system, but can dead nerve cells be 'woken up' or renewed? Taking human body cells back to their early stage when they are flexible enough to take on new roles would be a great biotechnological breakthrough, enabling the repair of damaged spinal columns or the growing of new organs. The lame could walk again, damaged livers could be repaired, and those with heart disease could be given a new lease of life.

Transplanted organs have a high risk of rejection, but not so a new heart grown from our own embryonic stem cells. These are found in abundance in the placenta we were born with, and some researchers recommend that parents arrange to have some stored for their children, just in case they are needed in later life.

Perhaps lessons can be learned from crabs, which can grow new claws, or the humble sea squirt which devours its own brain as soon as it attaches to a rock. It no longer faces the complex challenges of moving around, so it uses its own brain to provide the nourishment for the next stationary stage of its life cycle.

Our body cells do not have this flexibility: we cannot grow new fingers. Our sex cells have career options, but our somatic cells are given a job for life, bequeathing us an irreversible body-shape. This is the price we pay for having such a highly organised nervous system. We are not like axolotls that can grow new brain cells on demand. Nor are we like starfish, simply growing new parts when cut in two. Our cells are far more complex than five 'feet' surrounding a central stomach with a primitive proto-brain.

We are increasingly learning how even stroke-damaged brains can re-assign important tasks to other areas-in-waiting. Also, sufferers from Tourette's syndrome, obsessive compulsive disorder, depression and addiction can be taught a degree of self-direction through cognitive behaviour therapy, not just altering the synapses, as happens in the laying down of new memories, but something more fundamental: altering brain circuitry, forming new networks and remapping the brain.

They are aided in this by the discovery that, within limits, the brain is not fixed for life, or stuck in over-worn neural grooves, but capable of growing new neurons, thereby opening pathways less travelled towards a different future. The psychologist William James foresaw this possibility as early as 1890, and the neurologist Charles Sherrington described the brain in 1917 as an 'enchanted loom' capable of weaving ever-changing patterns.

Unfortunately this empowering vision was obscured for nearly a century by a deterministic approach to the brain, but neuroscientists and therapists now work much more closely together to find ways for the mind to redesign parts of its brain.

Time to settle down
Our brain maps not only overlap, but often drift or bleed into each other's territory, changing their contours on a daily basis depending on what we get up to. Neural pruning when we are young usually ensures that brain operations are ring-fenced, but for about four per cent of the population, the connections between the areas for analysing shape and colour stay unusually strong. As a result, synaesthesia or sensory cross-over continues into adulthood.

Synaesthetes retain some of the young brain's plasticity, scenting colours, tasting sounds or feeling them on their skin. A few go further, attributing personalities to numbers, and seeing auras around people. In what is known as mirror-touch synaesthesia, some feel the sharp pang of another person's pain when they see someone injured or in emotional distress.

These out-of-the-ordinary experiences can by turns be unsettling or the door to creativity and insight. Most of us lose this fluidity as we age, because an undifferentiated flexi-brain whose plasticity never solidified would have no starting point for building knowledge about the world, and openness to change has to be tempered by the need for continuity.

On the other hand, many brain cells remain fluid or uncommitted, which does not mean they are idle but on stand-by to cope with the odd and unexpected, ready to be mapped in, making the mind a complex mix of freedom mingled with restraint.

The process that triggers our brain cells into action is called learning, or the strengthening of synaptic connections through repetition of stimulus, starting in the womb and speeding up the moment we appear in the world. Through a ratchet mechanism called the Baldwin Effect, our brain has evolved genes for plasticity, enabling us to build new learning exponentially onto old, increasing our overall 'fitness' in the process.

We are born with the ability to respond optimally to whatever environment we find ourselves in, making our new learning almost instinctive, hard-wired over countless generations, and transmitted through cultural learning. All we need is inputs, and the more the better.

Some of our neurons are strongly genetically coded, giving us what we might call our instinctive behaviours such as rooting for the nipple the moment we make our appearance in the world. Our eyes merely have to see the light for our visual cortex to leap into action. Other neurons are experience-expectant, eager to reach out to their neighbours. The more we are bathed in love and language in our early weeks, the denser our synaptic canopy grows as our brain busies itself extracting the patterns of the world from the chaos of its inputs.

In our first two years we have to draft thousands of encyclopaedia entries about things, people, probabilities and appearances. The nature and quality of our experience will fill our brain with thoughts and turn our neurons into a mind. Most of our neonatal neurons will be free agents, highly associative and better than any other cells in our body at learning from their neighbours. Around the time of going to school we can claim that our brains enjoy better connectivity than our parents' brains, our neurons desperate to make new contacts.

As babies, before our neurons settle down or are pruned back, we are all synaesthetes, which might explain why children's stories are full of trains that can talk and pigs that can fly. In our childhood, our brain is still busy sorting the world, and hasn't yet erected fences between discrete brain areas, which means the boundaries between domains are still very fluid. If we suffer severe brain lesion to one side of the brain, our surviving half can take over most of the duties, but this adaptability declines rapidly after the age of ten.

This degree of brain configuration could not happen so easily in an adult brain, because for a brain to work effectively, cells eventually have to be drilled to perform specialist roles. By the age of ten most cells will be assigned to specific duties such as vision, hearing, language and so on.

We Are More Than Our Brains

Less is more

A cruel fate awaits a brain cell if it receives no stimulation or finds no useful employment. By a process called neural pruning, it will be weeded out, possibly millions at a time. Microglia cells, the brain's nurserymen, cut out sickly or unused synaptic connections to keep the system efficient. In a process called neural Darwinism, whole networks can be selected or deselected, as if they are fighting to justify the garden space they occupy. There is a tight developmental schedule to keep to, and no room for slackers.

Programmed cell death is ironically the key to life, and the sculptor of an organism's eventual form. In the womb, a baby's hand is webbed in the early weeks. What makes it a human hand, not a bat's wing, is that 'suicide cells' in the membranes between the digits begin to die off to create the gaps between the fingers.

By the same token, whole brain areas need to be cleaned out once we start learning to talk, repurposed or readied for the next growing spurt. In the first year of life, we rely on neurons that are highly sensitive to subtle differences in the sounds of speech, which is why a French baby doesn't grow up sounding like a German. By the age of twelve months, this job of fine discrimination of sound is done, and the super-listening neurons aren't needed any more. They wither away, making space for neurons that can fast-forward us to the next important stage, the acquisition of semantics (what things are called) and syntax (how to put words together).

We can learn more than one language in our first few years, before our first-language neurons are switched off. Bilingual children manage this feat by having a whole area conserved for their second language. This does not go to waste in non-bilingual children, but is diverted to other uses. The motto of the brain is, waste not want not.

By age ten however, when our brain is reconfigured for its next learning phase, this second area is fully colonised by other learning, making it much harder to learn a second language as an adult, and to sound like a native speaker.

Following a principle called Hebbian learning, named after the brain researcher Donald O. Hebb, neurons that fire together wire together, but those that we don't use, we lose. Strictly speaking the 'firing' metaphor is misleading, because what happens at the synapse is as wet as it is dry, but the mantra has stuck because it is catchy and makes the point so well.

We Are More Than Our Brains

Donald O Hebb
1904-1985
Hebb's motto was, 'Neurons that fire together wire together'. The more we use a part of our brain, the stronger it gets, because the connections between neurons are tightened.

Although we steadily lose neurons as we age, the ones that remain form much tighter networks across areas that have been worked hard. The paradox of neural pruning is that less can be more. Some psychologists suspect that autism, a mind condition emerging during the second year of life, characterised by impaired emotional development, might result from too many cells at the lower input level of the brain being left to suffocate higher executive regions, which need space and dominance to do their job effectively.

Once the structure is in place, efficient learning is not about how many neurons we possess in our 'grey matter', but how their connections are coated with white myelin to speed up message flow and memory retention in our 'white matter'. If we are to learn anything at all, millions of our neurons must lie in wait, unassigned or empty, until they are written upon by experience.

Think of all the skills we have to master before we can read: sight recognition of the alphabet, upper and lower case, how the letters combine to make words, what these words sound like, and all the exceptions to those rules. Consider the variations of pronunciation of the 'gh' sound in though, through, thought, tough, thorough, and trough. As well as the convention of what a book is for, and how print is laid out on its pages, we have to learn how sentences work, and how texts operate. There is nothing 'automatic' about these things, because they differ from culture to culture.

We Are More Than Our Brains

The learning brain

To learn something is to change our brain, either by rearranging what is already there, or 'filling up' empty space, creating new synaptic networks in the process, spread on a 'use it or lose it' basis right across the brain. But we never run out of space. The brain is so vast that we can learn new skills in later life, though they will take us a little longer to master. By a conscious effort, we can dissolve and remake many of our connections, otherwise we are stuck with what we know, never able to break bad habits. Learning is therefore reversible, and can be a lifelong process. This is a message of hope, especially for victims of trauma and drug abuse: what can be learned can also be unlearned.

Learning is not merely a cognitive reflex that happens when we are confronted with something new, or plonked in front of a screen. If that were true, our eleven years of compulsory schooling and a lifetime's television watching would turn us all into geniuses. Learning involves the whole organism, as an embodied mind, not a detached intellect.

As we grow, our emotional circuits are pruned or strengthened in the same way as our cortical ones, so it matters that we feel loved at home, shown what is valuable in our culture, and cared for as a member of a learning community. In what is known as neuroconstructivism, our genes, neurons, networks, emotions, bodily awareness and sense of social belonging form a ladder. The higher we climb, the more deeply we understand. The more we share our learning with others, the more sense we make of the world.

The growing brain and its plasticity are the playthings of both nature and nurture in equal measure. Experience is dynamic and unique, because all developmental pathways are open at birth. But it also leaves indelible footprints, because pathways cannot stay open forever. A child raised in the jungle will eventually possess a different brain from one raised in a city, as learning experiences are 'canalised', or cut into the mind like streams that find the fastest way downhill. Once adulthood is reached, these streams are all but impossible to redirect. There is therefore truth in the saying that you can take the child out of the city or the jungle, but you cannot take the city or jungle out of the child.

The legacy of experience partly owes itself to the brain's need to ensure maximum efficiency by neural pruning and running a tight economy. Children lose millions of redundant neurons between the ages of seven and ten that aren't yet wired up to precise functions in other busy brain areas.

83

We Are More Than Our Brains

Neurons are too precious to be left to lie idle or go to waste. Blind people colonise their unused sight areas to learn to read Braille faster, in a process called neural theft. Their fingertips become as sensitive to raised dots on the page as the eyes of sighted people are to printed text. In fact, without calling on the super-processing power of their visual cortex, made possible by cross-modal plasticity, they wouldn't be able to read Braille in the first place.

Like any muscle, our brain responds to demand. If we learn the violin, the connection swells between our fast-moving left-hand fingers and its control centre. But this dedication to a special interest can occasionally cause problems. With professional violinists and pianists, who practise for many hours, intensity of use can lead to a malfunction known as focal hand dystonia. The brain receives so much 'finger information' that it thinks it is dealing with one finger, not two, leading to loss of motor control in one finger, as if they are fused, requiring some difficult retraining of the brain to get both pathways working again.

This kind of therapy raises interesting questions about the mind's ability to modify the brain through the two-way process of neurofeedback. As well as performing finger exercises to reshape the brain through physical feedback, merely *thinking* about moving each finger separately can help to restore some function.

Physiotherapists give us exercises to wake up lazy or damaged muscles, and dance instructors get us to perform moves that almost make us lose our balance. These are hard at first, but they gradually get easier as our brain learns to listen to the messages from its body. This proves that the body as well as the mind can retrain the brain to get those limbs moving. When we break our leg, the challenge is not just to repair the bone, but to get our brain to 'wake up' its shrunken control area of that limb, and get it dancing again.

Neurofeedback has a psychological as well as a physical dimension. In the long term, each of us responds to trauma or copes with stress differently, or we can train ourselves to do so. In the short term, we can cheer ourselves up by whistling, singing, smiling or talking to a friend. This is very easy to put to the test. If we start laughing, at nothing in particular, we soon feel an improvement in our mood, which is why some of us go to 'laughter clubs'. Changing our emotion changes our brain, and vice versa.

We Are More Than Our Brains

Shaping and being shaped

This capacity of brain areas to shrink and swell on demand, discovered by neuroscientists but always intuitively suspected, has caused us to revise our view of our brain's neuroplasticity, or capacity to change. It is not a slave to its genes, but responsive to our experience: our brain assumes the shape of what happens to us, and what we pay attention to, moulded like a pebble on the beach by the waves of chance and necessity.

In other words, the life we lead and the meaning we make modify the structures of our brain. In an important sense our genes are not just automatic operators but are 'switched on' by experience. Through a process called epigenesis, nurture sculpts the brain as much as nature. Our life-world is moulded by society, and our biology is shaped by its environment.

Karl Marx would have been delighted that neuroscience confirms his keynote idea: not only do our minds make our world, but the products of our minding in the form of the 'superstructure' of society go on to make us in return. After a long period of strict Darwinian determinism, many biologists now accept that language and culture drive our future as a species as surely as our genes. Culture is as influential as biology in shaping both the species and the individual.

In the words of the playwright Arthur Miller, 'We are made and yet are more than what made us'. We form neural clusters and marshal our thoughts around the things we decide are worth looking at. Each thought and action lays down a few more memories and adds another layer of cells to our neocortex.

This means that the brain is always a work in progress, each thought shaped by those that preceded it, transcending but not forgetting what has gone before. This is not surprising, given that its watchword from an evolutionary perspective is adaptability, or its capacity to build on the lessons wrested from experience.

This does not mean that we can rewrite our genome, only our gene *expression*. Genes seldom work in isolation or as single-cause agents. As many as ten conspire to give us our eye colour, and although four thousand conditions have been linked to specific genetic defects, each follows its developmental pathway differently in every individual. Genes are therefore necessary but never sufficient to account for who we are and what we do.

Genomics is the study of how genes behave in whole organisms, rarely alone. For every gene that determines our height, we know through biofeedback that there is another waiting to be triggered or suppressed. A garden weed has a fixed genome, but its tallness and overall health is variable depending on whether its seed falls into fertile soil, or a parched crack in the pavement. We too grow more strongly if we are well fed and watered, physically, emotionally and spiritually, far less likely to fall prey to some of the ills lurking in our genome than our stressed acquaintances. We can't change our temperament, but we can learn to manage and modulate it.

The Zika virus alerted the world in 2016 to the dangers of brain damage in the womb, but that is not the whole story. Cases are on record of babies being born with only a rudimentary brain, then confounding the expectations of parents and neuroscientists by growing almost a full-sized brain in their early years.

It as if the brain has an overwhelming urge to auto-direct, desperate to grow the networks necessary to running the business of managing the body, opening up alternative pathways when others are denied. These cases of astonishing neuroplasticity are however very rare, and no such 'miracle' occurs in adult brains, where damage is far less easy to reverse.

We see similar indications of the brain's self-shaping power in the brains of the deaf and the blind. Their auditory and visual cortices do not lie idle but, in a process called cortical remapping, they enhance what senses remain. The blind learn to 'see' with their ears, developing more acute hearing; the deaf to 'hear' better with their eyes, enjoying more sensitive peripheral vision. Unlike skin or bone cells, neurons are not irreversibly programmed at birth to perform a single role, but offer a high degree of flexibility, because they are too expensive to be left unused.

But there is a price to be paid for this youthful neuroplasticity. In rare cases of the blind having sight restored in later life, seeing doesn't automatically happen, because so much of the visual cortex has by then been redeployed as extra hearing capacity, and it's too late to reverse this.

The teenage brain

As well as being vital to the brain's flexibility, neuroplasticity presents us with another paradox, especially to the teenage brain that goes through another hormonally-induced 'wiring up' burst at puberty. Psychologists have variously seen adolescence as a time of new birth, storm and stress, genital arousal or identity crisis, especially among males. Whatever happens, adolescence is not a disease, but a natural reconfiguring of the young into the adult brain.

Traditional cultures see the transition from boy to man, or girl to woman, as a managed rite of passage, involving a 'going out' as a child and a 'coming back' as an adult, having spent time alone in the wilderness, or under the tutelage of a mentor. Our ancestors did not have the luxury of a long life, and the transition from childhood to adulthood was quick. There was no time or scope for rebellion, petulance and delinquency, though there was always an element of challenge and risk-taking in the establishment of a new identity. Teenager-hood and all the angst that goes with it, such as status-seeking, cyber-bullying and body-shaming, is a modern construct.

Adolescence does not have fixed properties, our journey through it depending very much on the company we keep. Oliver Twist turns out differently from the Artful Dodger because he starts from a different place. A loving family home gives a very different kind of emotional security and education for life from a street-wise gang that can't be sure what threat tomorrow will bring.

The rule of thumb is that if children enter adolescence knowing they are accepted and valued, even if they kick over the traces for a while, they will emerge strongly on the other side. If nothing else, children have a psychological need to differentiate themselves from their parents, and this is bound to cause ructions.

However adolescence manifests itself socially, neurologically there is much going on. To make a butterfly, the goo inside a caterpillar has to be restructured. In the teenage brain this is achieved through heavy neural pruning, converting fast-thinking grey matter into the more deliberative connections of white matter in the cortex. This is prompted partly by biological programs and partly by experience, which increasingly takes over from the genes. The genes did most of their building work in the first ten years of life, but now life and learning have to take over, resulting in a more clearly defined sense of identity, and more secure decision-making.

The cognitive and emotional heavy lifting to achieve this can be done only after a substantial structural refit. Some have compared the teenage brain to a building site, with work going on from the age of twelve, and scaffolding still in place until the age of twenty four. The psychological challenge is that this rebooting must occur during a time of maximum exposure to the maelstrom of social media, making the adolescent world feel simultaneously exciting but bewildering, at once an exciting expansion and a necessary reduction.

The dynamism of the teenage brain is evolution's way of encouraging social change and cultural innovation through adaptation. Teenagers are the shock troops of new thinking, radical ideas and enthusiasm for new causes, but this inquisitive energy can just as easily be sidetracked into bad habits such as drug taking and antisocial behaviour. These become old tricks blocking the pathways to better solutions. With so many possible futures to explore, which adults have had time to evaluate, there are bound to be a few misadventures along the way.

Much to the concern of parents, the teenage brain craves peer group acceptance and social kudos more than worldly or material success. The flip side is that it is also motivated to take on grand causes and right the wrongs of the world. Adolescence is therefore a time of great energy, excitement and creativity, and if we lack passion during these years, we are not likely to find it later.

It is in the nature of the teenage soul to rebel, in order to establish an independent self. The challenge for those responsible for mentoring teenagers is to balance dull conformity with reckless adventure. Curiosity and idealism need to be nurtured, not thwarted, because by the end of adolescence, the habits of a lifetime are formed, be they iconoclastic or conservative.

Up to the 'tweenager' years, parents assume some responsibility for what films their children watch, how many hours they spend online, how much sleep they get, how late they are allowed out, who they mix with. They understand that, though teenage brains have the energy and performance of a racing car, they don't have the braking skills or road sense to match, or at least not yet. Unable to assess the value of what is worth storing for the future, teenagers need careful guidance and good role models. Only nurture and culture can provide the moral steering and qualities of mind that the impressionable young brain needs.

Eventually parents must let their offspring go, even though the brain doesn't reach full maturity until the early twenties. The second decade of life is a time of bewildering plasticity and emotional turmoil, a race to the finishing line between innocence and experience, novelty and familiarity, vulnerability and resilience, heightened risk and survival, confirmed by the high insurance premiums for drivers below the age of twenty five.

Societies manage the transition to full autonomy differently. Some recruit child soldiers as early as eight years old, others consider that minors cannot be held criminally responsible for their actions until twelve or thirteen. Children can do paid work at fourteen, ride a scooter at fifteen, marry at sixteen, drive a car at seventeen and vote at eighteen.

At the base level, our brain doesn't care much. It can't pay attention to everything all the time, so it opts for some stability in the chaos of all the flexibility. But it also needs the ability to rezone areas, repair damaged neurons, change its circuitry, break old connections and make new ones, otherwise we would be condemned to a life of stupidity, unable to change our minds when the facts change.

We are smart enough to survive the neural makeover of adolescence, fall in love a second time, weather grief, conquer addiction, learn new tastes and adjust to living in a new country. How else could we ever be able to quiet our minds with meditation, learn a poem by heart, or set off in a new direction? Even a bird can learn a new song each season, and vary its song according to its locality.

Changing our minds
Plasticity is essential, but so is stability. In personal relationships, this tension can be a boon. Some couples stay happily married for life because they still see each other as the person they originally fell in love with, while constantly remodelling their vision of each other in a favourable light. In the art and practice of life, flexibility and creativity are all.

It might be difficult to unlearn one skill and replace it with another, such as improving our clumsy two-finger typing with touch typing, but it pays in the end. In previous ages, we used to stay with one job for life, but today's volatile and rapidly changing employment market demands that we learn new skills several times over in our career.

We are fortunate that each neuron can have up to twenty thousand links right across the brain, the whole brain making a million connections a second with scope for a million billion more, which means that the brain's potential for change is practically infinite.

The implications of this for learning brains of whatever age are enormous: our brain's physical growth may be constrained by the size of our skulls, but its *potential to reconfigure itself* into new neural nets is vast. We could never exhaust our brain's learning potential, even if we lived forever.

This is a lifeline for sufferers from childhood trauma who have been blighted by abuse and need to over-write their pain with a more positive life-narrative. It also has important lessons for childhood education and lifelong learning. Unlike old dogs, we *can* learn new tricks.

Therapists are aware of the mind's capacity to heal itself after deep trauma, and teachers have known for centuries that minds are not fixed but capable of transformation and re-invention. Religious leaders have long understood the power of metanoia, or our ability to change our hearts and minds under spiritual guidance, and meditators have developed their power to adjust their consciousness, perhaps even their perceptions. Those in search of meaning can 'see the light', criminals can be reformed, and victims of cults can set their mind free from their controllers.

Neuroscience has exploded a stubborn myth that we are stuck with the brain we are born with. But we don't get something for nothing: if we are to take advantage of neuroplasticity, we have to keep our minds young.

Where does mind fit into the picture?

Physics – metaphysics – dualism – Descartes – monism – materialism – mind - consciousness

- Most neuroscientists are physicalists by nature, but accept as a working hypothesis that the mind-brain is a complex mix of physics and metaphysics.
- This is largely because a materialist or physicalist account of the brain is insufficient to account for the mind.
- An idealist or rationalist account doesn't work either, because it ignores how the brain works.
- Unless we combine physics and metaphysics, we cannot cross the metaphysical divide: how do we get from sense data processed by our brain to the lived reality of experience?
- This is not easy to answer, because no single area of the brain has been identified as the seat of consciousness.
- Buddhism's answer is to view consciousness as a series of distractions or 'unrealities' appearing on the screen of the mind, a theme picked up by modern philosophers of mind.
- Since consciousness's greatest trick is its ability to convince us that what we see is what we get, it makes more sense to view consciousness from an evolutionary perspective.

Going beyond physics

The discoveries of the neuron, synapse and neuroplasticity can be ranked alongside gravity, evolution and relativity as among the great advances in our knowledge of the physical world, except that this time the focus is on the hunter, not the hunted.

It is a giant leap however from the physics of the brain to the metaphysics of the mind, from the locations on brain-maps to the intentions of those who find their way by them, from the science of cognition to the art of being human, from the matter of the brain to the meaning of the mind. After several decades of searching for

91

their elusive prey, neuroscientists freely admit that tracking blood flow around the brain is the 'easy problem'. They have more data about neurons, networks and neural states than they can shake a stick at.

Consciousness however, or our amazing sense of the richness of experience, is dubbed the 'hard problem'. There is no 'theory of consciousness' that nails down how our mind 'makes' love, beauty, truth or spirituality real for us, and some think there never can be. Others insist we simply haven't been asking the right questions about the mind, or we don't yet have the right equipment to measure it, or both.

We have been quicker to catalogue the far reaches of outer space than the dark caves of our minds, because our intellectual journey outwards poses fewer existential challenges than our perplexing journey within. As we enter further into the age of machines however, we will need to cultivate the life of the mind more than ever.

We will also need to be clear about the difference a mind makes. While it is arrogant to suppose that animals do not possess mentality, intelligence and consciousness, we also have to recognise the ontological leap from animals to humans. We can appeal to nature for evidence of this, with no need for a supernatural creator: no other animal plans its future, manipulates its environment, passes on its culture or reflects on its existence in the way we do. A beaver can do these things up to a degree, but if we take it away from its watery life-world, it is lost. Our species has the versatility and technology to conquer every environment on earth.

The qualitative leap in the evolutionary process that led to Homo sapiens cannot be explained by Darwinian forces alone. Humans can fit themselves to any ecosystem, override their instincts, impose their will on their surroundings, fabricate worlds of ideas, form complex societies and shape the future, activities which require a motivated mind as much as a smart brain. But how and why did we make this leap? Philosophers call this chasm in our understanding of the nexus between brain and mind the explanatory gap.

Some claim that they possess the glimmerings of an answer, and it is a naturalistic one: mind arose for purely functional purposes. A mind minds just as a heart pumps blood. With endless time to get it right, and a large helping of serendipity, we can credit nature with the slow climb from a bacterium to a Bach cantata. Even a single cell can sense changes in its environment, and ten cells placed in a Petri dish seem able to self-organise into a connected network. This is the beginning of a biological brain and nervous system trying to integrate incoming data into a unified response.

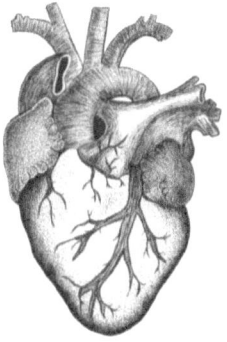

The heart is a pump
A mind minds, just as a heart pumps blood. There is no need for a creator, designer or ultimate goal. Even a fish has a heart, and for all we know, a fish has a mind too.

As inputs become more complex, and they are stored as personal memories, a kind of mind begins to emerge. This starts with the accumulation of data and the making of images, followed by their conversion into representations that symbolise inner and outer realities. The organism becomes aware it has personal experience in the present, can recall its past, and second-guess its future.

Perhaps we will never be able to explain fully how or why we evolved from single cells to complex organisms, or from hominid origins to Homo status. What we do know is that we find ourselves endowed with a new reality in nature. We can symbolise, think, judge, reflect, talk, care for strangers, create imagined worlds and take nature to pieces. We come equipped with a capacity to form a mind, no less real or evolved than our potential to breathe air or digest food.

Scientists debate whether the evolution of mind was pushed along by coming down from the trees, broadening our diet, upright posture, enlarged cranium, tool-making, cooking meat over the fire, or our ability to self-domesticate through culture. Some of these developments might have been proximate causes, some ultimate. A receding jawbone for instance, reshaped by natural selection because cooked meat requires less heavy chewing, allows for finer tongue movement. We cannot say that it led directly to spoken language, though it was almost certainly a contributory factor.

We Are More Than Our Brains

Bones, flints and camp fires can be studied through what they leave behind, but symbols, words, rituals, art and reason leave no fossil trace. At best therefore we can but speculate about how our species became the possessors of a mind so rare.

Physics explains only so much. Matter and energy are of a different explanatory order from what it means to be a person sharing the symbols and values of a culture. Biology also has its limits. From an evolutionary perspective, blushing is hard to account for. Why signal our embarrassment to those from whom we are trying to hide our feelings? Also, music and laughter are pointless activities, a waste of vital energy. Or was it always the smiling, cheeky troubadour who got the girl? Either way, we enjoy them for their own sake, not as part of a struggle for survival.

Accounting for mind

All we can say is that certain mutations took place in the human brain that gave rise to mind. Like it or not, scientists looking for material explanations for non-material phenomena have to settle for the fact that mind is as real as the nose on our face, even though their methods of analysis do not allow them to probe its deepest recesses.

We are creatures that dwell in two elements, matter and mind, needing both to complete our life cycle. We have to navigate complex labyrinths of genes and culture, reason and faith, things and thoughts. We face both ways in twin realities, simultaneously bound and free, rational and emotional, neural and mental. We are physics and metaphysics combined.

This doesn't mean we are condemned to gazing aimlessly at our navels. On the contrary, we have problems to solve and a life to live, so our minding has to be practical. Our brain evolved as a gap-filler, desperate to avoid ignorance, uncertainty and confusion, whereas our mind evolved as a 'social early warning system' to get us through to the next day without killing each other for precious resources.

As human evolution complexified, mind became an existential reality for a creature trying to solve the riddles of a perplexing consciousness and a challenging social life. Mind may well be an effusion of the brain, or an evolutionary quirk, but without it we can't have a private thought, theory of mind, social contract, life purpose, epiphany or sense of transcendence. Without a mind our language is an empty vessel, with no-one to learn, yearn, know, believe, reason, understand, argue with or apologise to.

In other words, we're stuck unless we grant that we have evolved to need two types of meaning. As well as wanting

94

material explanations for how the stars revolve in their courses, we also want to understand our human condition. We don't just want to know how things work, we want to know how to live. In fact this is the *first* thing we want to know. We are more likely to tell each other a story than recite a formula. But metaphysics remains hard, even for a metaphysician. In using our brain to understand our mind, we are trying to pull ourselves up by our own cognitive bootstraps.

We are reminded of M C Escher's sketch of a hand emerging from its frame to draw itself: the final picture is eternally doomed to incompleteness. We are left chasing our own shadows, lost in a paradox: if our brain were simple enough to understand, we would not be clever enough to understand it. But despite the metaphysical challenge, neuroscientists remain confident that the quest is worth pursuing, and teams of researchers continue to pool their findings about the most complicated organ in our body, and probably the universe.

The dilemma of dualism

Dualism, or the idea that mind and brain are somehow different substances, even if working in parallel, is instinctively disliked by neuroscientists, and yet they can't avoid it, finding themselves forced to use the double-barrelled label 'mind-brain' to describe what happens between our ears. There is a biological organ, and there is a sense of what it feels like to be me. But how to marry the physics with the metaphysics?

This is the challenge of the mind-body debate, which has ancient roots. Heaven-oriented religions have long portrayed the body as a mortal enemy, or temporary encumbrance that prevents the soul from flying free, and yet such a view might be little more than a consequence of our inherited sensory apparatus and limited cognitive development.

In our early years we experience the world as a unity, our sense of 'me' in a body that is separate from the world appearing only when we develop language which enables us to detach ourselves from our experience. We have to learn how to integrate 'me' with 'my body', otherwise we grow up feeling psychologically split, uncomfortable in our own skin, our body a stranger to us.

Our doctor has to confront the mind-body dualism when we become ill. If we report in with stress, are we afflicted with a malady of the mind or an imbalance of our body? Most therapists work on the basis that the mind resides not just in the head but is distributed through our whole frame, which is why stress might manifest as a skin complaint, stomach ulcer or hair loss.

Morning sickness for a pregnant woman is a case of her body knowing better than her mind. A sudden craving for one type of food, or aversion to one that was previously liked, might be her body's response to the growing embryo's nutritional needs, protecting it from potentially harmful substances.

Philosophers have poured more ink into mind-body dualism than any other topic. Four hundred years ago René Descartes thought he had solved it, boldly declaring 'I think therefore I am'. He saw his body as made of extended 'stuff' that it was the business of science to explain. Mind, the reasoning part of him, existed in a realm outside time and space, made of different 'stuff' altogether.

On the face of things, this is a neat solution to the age-old mind-body conundrum. It gave a boost to the new empirical science of the seventeenth century, and allowed religion to retain its jurisdiction over the eternal soul. It's also possible that Descartes was wary of incurring the wrath of the Church by being too materialist in his declarations about the soul.

Not all philosophers have subscribed to dualism, or agreed with Descartes. Shortly after him, Gottfried Leibniz remarked that looking for the mind as a separate entity was a bit like going into a flour mill, seeing the grain emerging from the grindstones, then asking, 'Yes, but where is the mill?' The mind, like the mill, is the *concept* that binds all the parts. Conversely, looking at a part of the brain and saying '*That's* where the mind happens' is as misguided as reducing the complex machinery of the mill to the churning of the grindstone, or the packing of flour into sacks.

This doesn't mean that categories such as mill and mind are less 'real', or not useful. We need them to organise our thinking at a higher level, and they are substantial in the sense that they allow us to create hierarchies of meaning, without which we could not ask this sort of question, or ponder the answer.

Whatever Descartes' original intention in dividing mind and body, Cartesian dualism is rejected by most neuroscientists because it leaves us uncertain who is the lord and who is the lackey. When we factor in Newtonian physics, the other great breakthrough in seventeenth century thought, mind and consciousness are stranded in no-man's land, leaving materialism and reductionism to pick over their bones.

We Are More Than Our Brains

The challenge of materialism

Tracking blood flow round the brain is one thing, but accounting for the miraculous growth of a newborn brain into a fully conscious adult mind is another. Evolutionists can't explain why the brain came to be as big and complex as it is, or why it is far more sophisticated than it needs to be for mere survival. Nor can neuroscientists reveal how consciousness works, only vaguely point to where it might happen in the brain.

Even so, some 'hardline' materialists insist on explaining consciousness as an inevitable 'emergence' from a brain with ten to the power of ten neurons, and ten to the power of fourteen synapses. If we put that many people into a crowded room and set them talking (it would have to be a room the size of the universe), it wouldn't take long for a complex society of ideas to emerge. Consciousness is the host who taps a spoon on a glass, calls them to order and extracts the gist from what they have all been chattering about.

Materialists try to 'explain' consciousness as an illusory outcome of neuroprograms going about their business, not immortal 'soul dust' but an evanescent sandstorm. To them, we're about as conscious as our computer is, because our brains work to similar algorithms. Once we've replaced ourselves with artificially intelligent machine-minds, which some in Silicon Valley assure us is just a matter of time, we can become pure brains or thinking engines, freed of all distractions. Until that great day comes, we are just brains on sticks, or the blobs that show up on our brain scans.

Seen through the lens of objective materialism, subjective experience is reducible to neural events and computational architecture operating in a global workspace. Consciousness is only the tip of what goes on in the chambers of our mind, while 'idiot programs' or 'default mechanisms' run the show below. Men for instance can have up to three erections a night in their sleep without even realising it.

The 'I' that Descartes claimed was doing his thinking for him has wilted sadly under the sustained pressure of materialism. There is no 'I', because there is no self, homunculus or little man inside our head watching the screen of reality and bringing everything together. There is only biochemistry and neural networks doing their job. Literature, art and personal experience may present us with an endless parade of 'I's' having lots of private thoughts and personal feelings, but these are beyond scientific investigation. No Newtonian theory of the 'I', the self or consciousness can be generated from reading people's secret diaries.

Standing up for consciousness

However we account for it, consciousness remains elusive. Since ancient times, it has been seen as something of a curse. Having disobeyed the gods, our punishment is that we are burdened with the awareness of the one thing we do not want to know: we are going to die.

Even in modern times, we have an ambivalent attitude to the defining feature of our humanity. Some see it as an illusion made by fireworks in the brain, though we could just as easily see it as our crowning creation. From an evolutionary perspective, consciousness is an unlooked for luxury, but not if we see it as a public show as well as a private theatre. Darwinian selection might have favoured those who knew their own minds well enough to read the minds of others.

For four centuries of materialist science, the standard story has been that only matter is real, pushing consciousness into the wilderness. This thinking may however be on the cusp of a paradigm shift. New ideas about energy, information theory and research into artificial intelligence are slowly challenging this orthodoxy, potentially turning dualism on its head.

If our mind can be downloaded into virtual form, and everything can be reduced to information, this means that there is a world of ideas that can exist independently of matter, regardless of the code it is written in, or whether it is stored in a human mind or a digital cloud. Matter no longer reigns supreme, and causality can flow top-down, not just bottom-up. The universe is minded, if only because we are here thinking the thought, and everything in nature sits somewhere along a continuum of consciousness.

Panpsychism takes this argument to extremes: even stones have minds. This goes too far for some, but once we progress from mineral to animal, even vegetable, all life forms show elementary forms of awareness and intelligence. Sundews can count, not releasing their trap until the sensitive hairs on their leaves have been tickled three times, making it more likely that an insect is caught.

Seen this way, consciousness co-evolves with life, and may even be an inevitable outcome of evolution. If Homo sapiens expires, another species will take our place, ensuring that the universe remains minded. This means that it is futile to try to explain consciousness by reference to anything else in nature. It is its own thing, a holarchy at the top of the hierarchy.

The phenomenal mind

The privateness of consciousness makes it almost impossible to study directly or objectively. In the early twentieth century, a group of psychologists who called themselves phenomenologists tried to get round this by concentrating on how sense impressions appear to our mind, often contrarily but no less 'real' for that.

Phenomena are 'things shown', and reality shows itself to us in tricky ways. Consider for instance our understanding of time. To a physicist it is constant, arrow-straight regarding cause and effect, but reversible in terms of the revolution of planets and possibly the evolution of the Universe, the Big Bang one day folding in on itself as the Big Crunch.

To the conscious mind however, time is a one-way ticket to the grave, running slowly or quickly depending on what state of consciousness we are in at the time: wakefulness, dream sleep, deep sleep, daydreaming or heightened awareness, however induced. During orgasm, our awareness of time switches off altogether.

Time seen, felt or measured from the inside by our mind is an entirely different phenomenon from what our brain reads off a chronometer. No wonder a watched pot never boils, and an hour waiting for a train feels longer than an hour watching our favourite movie.

Time expands and contracts like this because it is a series of lived moments, inaccurately recorded as memories, fleetingly experienced in the present, dreamily hoped for in the future. As a result we end up with two 'times', or two 'reals', one 'clocked' on the outside, the other 'lived' on the inside.

Phenemonologists, paving the way for cognitive neuroscientists in a later generation, concluded that our thoughts don't just happen, or bubble up randomly from below. Our mind is constantly reaching out to things, so our thoughts are always *about* something, responding to our surroundings, attempting to create fixity from flux. It's not so much a question of what consciousness *is*, but what it *does* for us: we are what we think about.

This 'awareness' is never easy to pin down, because consciousness doesn't just suck in stimuli like a sponge: it selects, mediates, interprets, prioritises. In that sense, we are what we pay attention to. Support for this idea comes from a surprising quarter: quantum theory. Inside our brain there are thousands of possible worlds, but our consciousness collapses them into this one, here, now.

We Are More Than Our Brains

If not, we would be overwhelmed, incapable of choosing whether to eat an apple or a banana. We have to step into the stream, and when we do, we alter the flow. When we wake up in the morning, there could be a million more versions of us across the multiverse, opening their eyes a microsecond earlier or later, rolling out of the other side of the bed, putting on slightly different clothes.

Reducing infinite possibilities to singular probabilities does not always work perfectly. The insula and temporoparietal lobe play vital roles in integrating inputs in the brain, but when this process misfires, we might end up feeling there is more than one of us, experiencing an out-of-body moment, suffering from multiple personality disorder or hearing voices in our head.

All creatures need to know where their body begins and ends, otherwise they might end up eating themselves. For humans, a degree of flexibility in establishing body boundaries extends our sensory range and enriches our consciousness. The angler's focus of attention can be many metres away, at the end of the hook, travelling up through the slightest vibrations in the line and the rod. Empathy can put us in someone else's shoes, and in the future we might find ourselves travelling strange new worlds in the body of an avatar.

The existential mind

The next thinkers to stand up for consciousness were the existentialist philosophers. Starting from the premise that we have no grand meanings to fall back on, we are obliged to create our own. We are forced to be free, 'thrown' into the world, so we must create a consciousness that allows us to feel at home in it, not alienated. Every decision we make defines who we are. By focussing on felt experience we put ourselves back in touch with our bodies, and make ourselves aware of the uniqueness of our life-world.

This is not an invitation to solipsism, or cutting ourselves off from others, though existentialists are sometimes criticised for underestimating the contribution of the wider world in which we find and make meaning. They overlook how strongly we are connected to our history and culture, which present us with moral obligations to each other.

How we respond determines our authenticity or insincerity. Existentialism appealed strongly to French resistance fighters in World War Two, many of whom felt that collaboration amounted to false consciousness. To resist was to be fully alive, even though it risked death. For many Germans, it meant committing to

Nazism, though a few spoke out against it. Existentialism asks us to make a decision, but it does not guarantee we will be guided by reasonableness or compassion. For that we must look elsewhere.

Georg Hegel
1770-1831
Hegel did not see mind as an evolutionary afterthought but as the driver of human history. His notion of a 'world mind' is too vague for neuroscientists wedded to empirical enquiry.

Two centuries earlier, the German philosopher Georg Hegel had cast the net of consciousness even wider, suggesting that the history of the world is a gradual coming-to-awareness of the human spirit, or a coming-to-consciousness of the universe itself. Some believe that Continental philosophers, by showing how inner experience can integrate the mind and the body, and be objectified to a degree, have been more successful than Anglo-Saxons in resolving Cartesian dualism. Others accuse them of being too mystical, ignoring the dictates of biology.

More recently, psychologists, neuroscientists, even quantum physicists, have come to see the mind from an existential perspective, not just as a hallucination or trick of language. They have begun to grant an active role for consciousness, even though they can't explain it. They have accepted that mind, through attention and intention, can exert real physical effects on the brain. Mind does matter, and matter is not mindless.

A choice of bookshelves

We see the legacy of Cartesian dualism when we go into a bookshop. When the eighteenth century philosopher David Hume perused his bookshelves, he divided their contents in two. There were those books founded on hard, verifiable, empirically observed fact, or what was later called 'positivism'. The rest he dismissed as unfounded speculation or sophistry, worthy only to be thrown into the flames.

He was being deliberately provocative, but his classification is still clearly visible in a modern bookshop. If we want books on the brain we must go over here to the science section: the brain belongs to the hard-nosed worlds of physics, biology and chemistry, replete with cells, axons and neurotransmitters, all of which can be seen on a scanner. Over there is the touchy-feely 'self-help' mind and spirit section, the realm of metaphysics, beliefs, psychotherapy, culture and society, which belong to experience.

We seem forced to choose: do we want to know about alpha waves and the neocortex, or mindfulness and the meaning of life? To keep their job manageable, neuroscientists don't fret too much over this mind-body dualistic puzzle that has kept philosophers busy for three thousand years. They keep it simple by sticking to physics, ditching the metaphysics and declaring themselves as proud monists: there are not two substances but one.

Mind is brain. Genome is to body as connectome is to mind. What we experience as mind is simply brain states and neural substrates. We are our brains, and our brain is the locus of all we know, who we are and what we hope for. Joy and despair might be *experienced* as intense feelings, but to brain science they are *processed* electro-chemically through relentless cause-and-effect mechanisms in specific neural networks which show up on a scan.

But as the mind-brain's awkward double-monicker reveals, monism takes us only so far. It doesn't chime with what being human *feels like*. We are brain *and* mind, making us natural dualists whether we like it or not. We are victims of our own brain's evolved 'binding' capacity to pull all its operations together so that we see ourselves as a unity of experience, not lots of different neural networks firing away mechanically and separately.

In that sense, from the point of view of lived experience, neuroscience works back to front. When we're stressed or trying to make sense of things, we don't go to the science bookshelf to learn the label of the brain area that is controlling our mood, or the name of the neurotransmitter that is holding us in thrall. We go to the self-help section to benefit from the advice of others who have weathered the storms of life, or learned the lessons of mindfulness.

Seeking a middle way

There has been a growing acceptance in second-generation neuroscientists that while we possess a biological neurome or brain, experience wires this into a personalised connectome or mind. Evolution had to bequeath us not just a high-maintenance neural processor to make sense of the world, but also a mechanism for establishing our place as an individual in an intricate social web.

In early neuroscience the traffic was always one-way, from brain to mind, but there is growing experimental evidence and justified belief that there is a steady contra-flow: the mind can affect not just the body but also reconfigure the brain, a truth intuited by Buddhism two and a half millennia ago.

In this new thinking, the mind and brain don't sit on separate bookshelves, but populate a shared biological 'field' or life-world. A brain-in-a-vat is a fine thought experiment, but it cannot exist in reality. Just as a bird needs air, wings, nest, eggs and a place in the pecking order, so a brain needs a body, sensations, thoughts, felt experience and lived history. In that sense, the brain is as much a sociocultural artefact as a biological organ, evolving not just as a survival machine, but as a member of a community of minds.

Nevertheless, some neuroscientists are 'strong brainer' monists, or single-substance theorists: the mind-brain is made of one stuff called matter. 'Weak-brainers' are dualists, believing that mind generates a level of reality that is ontologically different, giving us the quality of being-in-the-world, not explainable by physical brain science alone.

The good news is that in normal life we don't need to sweat the philosophy too much: we have evolved to experience each moment as a whole person, never as a female, psychopathic, rational or religious brain. Our sense of *being here now* is indivisible.

Neuroscientists tend to dispense with metaphysics, which at least helps them to focus on the physics. They work in the laboratory with an MRI scanner nearby, intent on deconstructing the brain for medical or technical purposes. It is much more difficult to infer from the blobs on a scan the thoughts and feelings of the person inside the scanner, or how the mind evolved to process the world in any particular way.

We make choices based on values which are forged outside the laboratory. They are not written on the surface of our cortex at birth, and not all of our psychological problems are reducible to scientific solutions. Brain science is a powerful discipline, but it cannot explain all our complicated metaphysics and philosophy as just so much biology and neurology. Consciousness cannot be reduced to a multiple-draft fitness-maximising machine.

We Are More Than Our Brains

Mind-brain dualism might be unavoidable, an essential condition of human minding, suspended as we are between thought and feeling, which is why we are obliged to use an awkward hyphenated compound. Despite the fact that the mind-brain dances as seamlessly as a couple who mirror each other's movements perfectly, there is no single word in our language to cover the beautiful pas de deux in our heads of brain and mind, of head and heart, of biology and consciousness.

Mind alive

We know we can't live without a brain: 'brain dead' is the final pronouncement of doctors before they switch off our life-support machines. A dead brain showing no vital signs is a spiritless machine that cannot house any kind of mind. If we jump off a cliff, we kill our body, brain and mind all in one go.

It's not so clear however what happens to our mind if we survive the fall, but end up in a coma. Our brain is still alive, but without awareness of what state we are in. Without thoughts flowing through it, and awareness of them, our brain is an empty vessel, containing no conscious mind. If we emerge from the coma and regain consciousness, we get our mind back.

This suggests that mind adds something vital to brain, a truth Shakespeare captures in his phrase 'the marriage of true minds'. We mind how we go and we bear things in mind. Brain and mind are two sides of a coin, indivisible in practice as a mind-brain: we can't peel heads from tails. Being brainy and being mindful are cognitive complements, weighed in the balance of experience.

Just as wetness can simultaneously be water and H^2O, so consciousness can be at once a material property and a subjective awareness, a function of molecular traffic and a web of symbolic significance. We can quench our thirst with H^2O or be refreshed by the balm of Gilead. Depending on our perspective, we can be cheered by a familiar face or zapped by Prozac: the effect on our brain chemistry and our 'state of mind' is the same.

Though we have as yet no unified theory of brain and mind, we do know that they cannot be existentially separated. Degenerative brain disease makes us more aware than ever of the brain as the hard disc of our identity, but feats of human endurance and positive thinking remind us of our ability to rewrite our mental software by practising mind over matter.

So while we can justifiably see our brain as our *necessary* biological survival system and fount of our subjective experience, it is not *sufficient* to characterise all our minding as kinds of brain activity, reducible to the atomic or molecular level.

104

Unifying consciousness

The unifier of the mind-brain is consciousness. Conscious*ness* as an abstract noun sounds vague: where would we start looking for it? It feels more real and accessible if we reframe it as actions such as looking, judging and deciding. Even in our dreams, we are always *conscious of* something in particular, of having something *on our mind*. We can't discuss anything about the mind or brain unless we have *in*tention as well as *at*tention. In other words, we must make a conscious decision to do so.

Regardless of how consciousness evolved, or whether we explain it as 'global workspace', 'integrated information', 'bicameral mind' or 'phi' (how tightly connected our neocortical neurons are compared to other animals, or even each other), consciousness is a whole-body experience: we feel it in the tingle of our fingers and the tickle of our toes. On the other hand, we know it resides largely in our brain, because people who are paralysed from the neck down are still fully aware of the succession of their mind states.

Being conscious is not therefore an optional extra. It is an irreducible feature of being human, a guarantee of our participation in the world. To think is to be conscious, to be conscious is to think. Whether we believe we can explain consciousness neurologically, or insist that we can never have a comprehensive theory of consciousness because we can never escape the loop of our own thinking, we still have to perform a conscious act of mind if we are to make any headway in deconstructing a working brain.

Brain registers the sound of music, but our conscious awareness of 'being here now' responds to its charms. There's not much point listening to music while we're not of a mind to enjoy it. Also, music vanishes on the wind, so any lasting meaning it has for us has to be an act of mind and personal memory. Brain can be objectively seen, as a multi-coloured MRI scan, but consciousness is subjective experience, hidden from view.

While I *use* my brain and *show* you my neural printout, I *get to know* my own mind and *express* my feelings, which bring the coloured scan to life. My illuminated brain is a biological organ, but my mind is a different kind of light show, made up of insights into the consciousness of self and others.

So while we may study the brain as 'it' from the outside, we have to live our mind as a conscious 'I' from the inside. This gives us a sense of living in the first person, in what has been called 'the presence chamber of the mind', or 'feeling tone of intimacy'.

We Are More Than Our Brains

Consciousness could not evolve in isolation, as a front row seat in a private theatre of experience. It also needed to be shaped by engagement with other minds, in a socio-cultural public performance as 'real' as our bones, skin and teeth, shaped by the thoughts we think, and the words we use.

This does not mean we have open access to other people's minds. We often misread their meaning or intention. Nor can we guess what sort of consciousness Egyptian slaves enjoyed as they toiled on the Pyramids. We can only assume that they had no mental space for our modern notion of individuality, and were not likely to burst into a chorus of 'I want to break free'.

We know that consciousness has a bandwidth, which some of us can control better than others. We use broadband when strolling through the park and taking in the scene generally, and we switch to narrow band when we see something that grabs our attention. We might have several channels running simultaneously, such as when we daydream, meditate or watch a movie, not to mention take a mind-altering substance.

Deciding how far animals can claim to be conscious has proved to be controversial. All biological animals possess survival mechanisms called brains, and we know they feel pain, but that does not mean they are conscious in the way we are. And yet, although we see our dog as 'our dog', our dog might be sitting there thinking 'that's my owner'.

Thomas Nagel wrote a famous essay entitled 'What it's like to be a bat'. His point wasn't that our lack of sonar prevents us from 'seeing' what a bat 'sees'. We don't even possess the mental structures to imagine what consciousness must *feel like* in bat-world. These are important considerations to animal rights campaigners, crucial in deciding what suffering an animal such as a chicken might undergo while cramped in a cage in a factory farm.

Some are convinced that their pets possess minds and consciousness, because they see evidence of high levels of intelligence, decision-making and emotional response. The more we learn about animal brains, especially the higher status-seeking mammals which live in complex groups, the closer we realise they are to us in their minding. They simply possess different *kinds* of minds.

What the Buddha saw

It was human consciousness that fascinated the Buddha as he sat under the bodhi tree in India over two and a half millennia ago. He saw our mind as trapped in 'gross apparent reality', or the ignorance of the body, but it was spiritual illumination he sought, not scientific certainty. Desire is infinite, and unless we train our minds on higher things, we are doomed to bounce between pleasure and pain for eternity.

Buddha saw the mind as a restless monkey that plays tricks on us and creates its own reality, even in sleep, caught up in relentless sensory bombardment. The images that flicker before our inner eye are the confections of our mind, not 'real' or 'ours' in any meaningful sense.

This idea of inconstancy and flow, central to Eastern thought, had to wait two and a half thousand years before modern evolutionists began to see life as a dynamic unfinished process, never static, and physicists to see matter as a ceaseless round of particles randomly coming together and flying apart, never fixed.

This understanding of the mind as ephemeral flux, our thoughts flitting like birds across our mental screen, persuaded Buddha that nothing is permanent. The wheel of suffering spins us round in a whirligig of endless causation and insatiable desires. Our mind is an entertainment system of flashing lights, diverting us one moment and demanding to be noticed the next.

Our wants are passing fancies, and any sense of an 'I' in control is *maya* or illusion, a deception linked to the word 'magic'. Once one need is sated, we move on to the next in a futile round of satisfaction-seeking, like a child in an amusement arcade, distracted by noise and consumed by constant craving, never accessing the 'real' behind the appearance.

A science of the mind

Although he has been turned into a god by later devotees, Buddha was a philosopher, not a manifestation of divinity. He was not so foolish as to try to persuade us that we don't exist: we clearly do. His intention was to free us of delusion, by exposing and then eliminating the myths we live by.

To this end, though he could not lay claim to being a neuroscientist, he offered us a science of the mind, a kind of self-study of our own thoughts. The doctrine of dependent origination teaches that we don't think our thoughts, they think us. If we are angry, it's only because we permit ourselves to think angry thoughts. If we are offended, it is a fleeting emotional state that registers the hurt, not an inviolate and essential core self. If we ignore the insult, nothing in the world is materially altered. If we suffer because of it, it is because we allow our negative thoughts to overwhelm our positive ones.

We Are More Than Our Brains

If we feel unhappy with our life, it is because we are fixated on what we cannot have. If we feel guilty, it is because we have not understood how our brain has evolved to be addicted to stimulus, hyper-sensitive to change and over-anxious about the future.

Buddha is often accused of offering a counsel of despair, but for many he grants the power of self-enlightenment to give peace from incessant striving. Over two millennia before Freud, he urged us to get to know our mind, not just the noisy antics up top that force themselves on our attention, but also the annoying satisfaction-seeking cravings that constantly bubble up from below.

He had no intention of reducing us to the matter of our brain. Instead he wanted us to understand our emotions, not to be blown along by them like leaves in the wind. He wanted to point us towards a deeper appreciation of mind, not to be mind*less*, but to be mind*ful*, discovering tranquillity via exceptional mental control, a discipline the hasty West still finds little time for.

His ideas have been given new life in the neuroscientific findings about neurofeedback. The more familiar we make ourselves with the machinations of our brain beneath the surface, the better chance we have of laying claim to self knowledge, or controlling our next move. We can turn our negative illusions into positive insights.

The Buddha
The Buddha, living around the fifth century BCE, cautioned his followers against distraction brought about by a constant stream of sensory inputs.

Few of us have the patience to become practising Buddhists: life is simply too busy. Nevertheless, 'mindfulness' has taken off as a kind of diluted Buddhism adopted in the West as a form of self-help or social transformation. Even business executives are urged to be 'mindful' towards their employees. Few grasp that

We Are More Than Our Brains

Buddha also taught a much more difficult doctrine of *anatta* or no-self. This is a counsel of self-discipline, not the egotistical self-realisation that many turn to as an antidote to the febrile distractions of modern life.

The Buddhist notion of the unreality and impermanence of the self seems to foreshadow another core teaching coming out of neuroscience: the self is an illusion, and the mind is 'flat', driven by hidden urges we never fully get to grips with. The Buddhist concept of 'no-mind' does not however mean absence or shallowness of mind, but a new appreciation of what is 'real', achieved by a level of control that comes only with concerted effort.

No-mind also entails compassion, because once we have subdued our own desires, we can become more aware of the needs of others. Buddhism emerged from a much older Hindu tradition that taught depth, not shallowness: we are *atman*, born divine, a fragment of a deeper reality called Brahman, and our life's task is to find our true self within this eternal scheme. Hinduism offers us Being, or the sense that we belong to a greater and lasting whole. Buddhism's departure was to preach Becoming, or the fitful activity of a brain defying the entropy and impermanence of an endlessly disappearing material world.

The mind-brain quadrant

It is the human condition to inherit a peculiar four-way legacy of a minded brain or embodied mind in a spiritual frame. William James, the father of psychology, was among the first to suggest that we are complex blends of the biological, the natural, the social and the spiritual.

The quadrant below shows how, even though our mind-brain can be compartmentalised for study purposes, it is held together in a still point at the centre which we call the conscious self, where we choose and express intentions as free agents, always in a social context. No quartile exists in isolation, because all four belong to Popper's World Three, which we discussed in the introduction.

The conscious self is not therefore a mere extravagance of an over-evolved primate brain, a figment of language, a neuromyth or a user-illusion, but a vital evolved capacity to navigate our way through a world of other minds, and to select what matters to us from an otherwise overwhelming range of options, without which we have no way of knowing who we are.

For many, the upper half of the quadrant is too metaphysical for consideration in the dominant techno-scientific paradigm of

our age, and yet, if we deny its reality, we create a metaphysical split that erases half of our experience. It is not a
contradiction to see ourselves as both sentient biological animals and sensitive human souls.

Mind	Spirit
psyche/anima	*pneuma*/breath
meaning maker	immaterial soul
reason	will
beliefs and values	aspects of eternity
Metaphysical divide	
Brain	**Body**
cerebrum/cognition	*sarx*/meat
machine	flesh and blood
logic	emotions
electrochemical networks	hormones and enzymes

The problem with diagrams of any kind is that they make us think that these divisions are real, like lines drawn across our cranium, when in fact the brain is a multi-mind, a continuum of connections and stream of awareness. Every single thought we have engages every quadrant.

Diagrams are two-dimensional, whereas brain/body/mind/spirit operates as a four-dimensional unity, lying not alongside but inside each other, as do culture and society. When we are ill, our doctor knows that many of our symptoms are psychosomatic, or signify a spiritual need, our physical and psychological elements running through each other like veins through marble. The true smartness from a therapeutic perspective resides in recognising that we are embodied minds and minded bodies, more likely to get better faster when all our parts are treated as one.

110

How do we get smartness from simplicity?

Brain size - vital statistics - intelligence - primitive brains - learning algorithms - metaphors of mind – reductionism

- Smartness isn't a function of brain size, but of how experience wires neurons up into powerful networks.
- There is no such thing as a 'primitive' brain. The remotest tribal groups possess the full potential of human intelligence.
- The brain's operation is a complex interplay of physics, chemistry and biology.
- The brain runs on algorithms, but AI engineers are some way off discovering its Master Algorithm, or what joins everything up.
- The brain is not a machine, a computer, a pump, a puppet or a quantum field. These are only metaphors.
- To get behind the metaphors, we can employ reduction as a useful tool to establish causes at the lower levels of a ladder of complexity.
- Reduction*ism* is reduction taken too far. It fails to acknowledge that new realities can emerge higher up the ladder, such as beauty, justice, self and free will, which cannot be explained backwards, or in terms of lower levels.
- The trick is to match the type of explanation we need to the level of complexity that is involved.
- Taking things apart can help us to see things whole.
- It can also increase our sense of wonder, so long as we also sense an underlying order and harmony at the heart of nature.

It's not about size
The big brain arrived late in the evolutionary story, and is possessed by very few species. It is an expensive investment, and for creatures that possess one, such as higher primates and dolphins, its evolution has been variously put down to broad diet,

111

complex social life, sexual selection, and a vague idea called emergence, which we discussed earlier. Perhaps it was a combination of all of these.

But intelligence and evolutionary success are not just a matter of size. One of our ancestors, Homo florensis, was half our height, and therefore smaller-brained, but human nonetheless. An elephant's brain is bigger than ours, with three times as many neurons, but it can't do algebra. Our brain is ten times larger relative to body weight, giving us the highest encephalisation quotient of any animal on the planet.

Dinosaurs had tiny brains, probably dictated by the laws of physics. Their bulk rendered their movements ponderous, and their nervous system took longer to get messages in and out. They also had no predators, so no pressure to evolve the smartness to escape.

Smallness of brain isn't intrinsically a mark of low intelligence, as we see in the purposeful activity of the ant. Its brain can cope with about thirty separate mental operations, and a colony of ants constitutes a giant collective brain. Nor should bird-brain be used as an insult, because small birds can navigate vast distances to find their way back to the very copse or upland meadow where they were born. Their brains may be tinier than ours, but they think much faster than we do.

A microchip is even tinier than an ant's brain, but it can contain far more information than our feeble memory allows. What creates human intelligence is not compactness, largeness or raw computing power, but connectivity. A computer can thrash us on processing speed, but it does not yet possess the creativity to think up and design machines like itself.

Neanderthals had bigger brains than us, but they are extinct, and we are not, though a few of their genes survive in our genome. They appear to have used more brain capacity for sight and movement and less for reflective thought. If so, it is clear which turned out to be the evolutionary winner. Homo sapiens had the edge because we uniquely inherit more high-powered neurons in proportion to body size that allow us to think not necessarily faster, but totally differently from other species.

Our sapiens brain capacity has doubled since our Homo forebears walked the earth six million years ago. It had to get bigger to fit in some sophisticated processing: handling tools, reading other minds, thinking symbolically, talking and passing on culture.

There may however be a high price to pay for this. Our big brain has made us the most successful species, but our swelling population, technological wizardry and ecological impact have also put us in great jeopardy. Our brain's analytical smartness seems to have outrun our emotional maturity. An adaptation that put us ahead of the game on the savannah might not be clever enough to save us on a crowded planet with dwindling resources, disrupted ecosystems and contested borders. The average survival time for a species is five million years, and in our present form we might fall far short of that.

Our cleverness might even be making us less smart. We have created so many comforts, protections and safety nets for ourselves that aspects of our intelligence have been deselected, our brain size shrinking by as much as ten per cent since our ancestors left Africa sixty thousand years ago, on a perilous journey in which only those with the sharpest wits survived. Intelligence thrives not on size but on challenge and hardship.

As a baby, our head looks swollen, already a quarter of its adult size. But it still has a lot of growing to do. It is ninety per cent of its adult size by the age of six, expanding the last ten per cent well into our twenties. To accommodate this ballooning, the bone plates of our skull do not finally fuse until then. This means that in an important sense we are born prematurely. Not only does our skull need to swell, there is also a huge amount of wiring up to do inside.

Gestation in humans takes about as long as other primates, but what distinguishes human brain growth is that from the get-go in the cradle we activate qualitatively different brain grammars that propel us beyond the here and now into a realm of abstractions. Parallel processing allows us to run several streams of thought simultaneously, escaping the prison of our immediate surroundings.

We can achieve this because we enter a world of words, without which we couldn't spend the first third of our lives absorbing the symbol-laden riches of the culture we are born into, or form the dense connections of memory, reason and emotion.

Our big brain cannot get much bigger, because our physiology sets a limit. Science fiction films often show humans of the future with swollen heads in relation to their bodies, to accommodate their enhanced brains, but no woman could pass such a monster through her pelvis. Also, our bodies would require radically different structure to support the extra weight, we would need a childhood lasting decades, and super-schooling to reach intellectual and emotional maturity.

113

We Are More Than Our Brains

Connected intelligence

Smartness is not therefore simply a matter of brain to body ratio, otherwise going on a diet would make us more intelligent, which it doesn't. Nor is it mere size. It's all about connectivity, or how our brain is 'wired up'. It's about architecture, not bulk. At birth a child's brain possesses the same number of cells as an adult, and even more synapses, but experience has yet to shape them into a unique individual. What makes a mature brain is its personal connections, not being a 'big head'.

There is a mild correlation between big brain size and high IQ, but we are all familiar with the paradox of the big-headed night-club bouncer and the small-headed professor. Napoleon was small in stature but also a high achiever. Geniuses aren't necessarily tall, and the brains of sufferers from dwarfism are fully functional. The brain of an endurance athlete or a pregnant woman suffers slight shrinkage, but this is because it prioritises other energy demands on the body during times of stress, after which it reverts to normal size.

The brain's remarkable capacity to self-organise means that even those born with abnormal brain structures can go on to develop normal mental faculties. Sufferers from the medical condition of microcephaly however, or small skull-size that constricts brain growth, generally face a lifetime struggle of cognitive impairment.

Einstein left his brain to be dissected for science, but it turns out to be no bigger than yours or mine, just uniquely wired. Female brains are on average ten per cent smaller than male but possess just as many neurons. They are not less 'clever' than men, but 'differently clever', as their hemispheres are more tightly connected. The key determinant of male-female brain difference is the impact of hormones, which is a matter of biology, not gender politics.

Primitive thought

Among all human groups alive today, there is no such thing as 'primitive thought' or a less evolved brain. Early white colonisers believed that the 'savages' they came across in far-off lands were lower on the evolutionary scale. Charles Darwin, an enlightened thinker on most topics, believed that the brains of the natives he encountered during his voyage on The Beagle, living in tough conditions at the tip of South America, were smaller than those of more evolved Europeans. He even suggested that exposure to 'civilisation' might boost their brain size.

114

We Are More Than Our Brains

Anthropologists who have spent time with tribes untouched by civilisation have corrected this view. When local conditions are allowed for, most discover a common humanity and intelligence in all known human societies, confirmed by modern genetics.

They have identified what they call 'liminal' thought, or a pan-cultural instinct to divide the world into categories such as sacred and secular, wet and dry, clean and unclean, the living and the ancestors. They realise that the argonauts who navigated vast expanses of the Pacific Ocean on precarious rafts were skilled in 'reading' the messages in the waves, stars, winds and clouds. These 'natives' were not scientists who understood nature's causes, but they were perfectly capable of grasping its foundations in patterns and correlations.

In the 1950's Claude Lévi-Strauss suggested that the 'savage mind' is in us all, and that the brains of Bushmen and Aborigines are capable of every modern function or thought. Their spatial intelligence is almost certainly superior to that of modern city-dwellers.

He also argued that we are wrong to dismiss myth as a primitive form of human thought. Its attempts to explain the world and human actions are 'true myths', the work of a rational mind, just as early twentieth century 'scientific' ideologies about race and IQ are irrational, based on 'false myths' about culture and heredity.

Nor is skull shape an indicator of qualities lurking within. Statues of the Buddha show a cranial 'bump' signifying enlightenment, and Victorian phrenologists believed they could identify criminality by tell-tale protrusions of the skull, but both beliefs have no basis in science. The brains of Buddhist monks and Victorian pick-pockets are no larger or smaller than yours or mine, their cognitive potential neither superior nor inferior, but wired differently by the dimensions of their life-world.

What sets us all apart is the unique configuration of our neural connections and thought-patterns, which are largely under our own control, so it's no use blaming our parents, teachers or government officials for our moral lapses or intellectual shortcomings.

Vital statistics
The natural sciences contribute equally to our understanding of efficient brain functioning. Biology accounts for sensory feedback, the hoovering up of dead neurons and replacement with new cells. Our development is powered by genes, few of which are unique to the brain, but the majority might be expressed in our neural activity at different stages of our life.

We Are More Than Our Brains

Genes are often described as forming a blueprint, but this is misleading: they are too responsive to feedback for this static metaphor. Nor are they in a constant 'selfish' zero-sum struggle for survival with each other, or even their owner: their watchwords are adaptability and cooperation. A complex organ like the brain could not function without a high degree of feedback and mutuality.

Chemistry tells us there are one hundred and fifty chemicals ferrying messages around the nervous system, not as a mineral sludge, but as a finely mixed cocktail out of which arise the emotional and intellective properties of the mind. Up to fifty chemicals are simultaneously active in the brain at any moment. Whether we are asleep, alert, moody or elated, our brain state is always under the influence of a particular chemical catalyst.

Physics tells us that electrical brain activity can be detected as early as the third month in the womb, an adult brain burning fifteen watts or the power of a dim light bulb while idling, up to four times more when concentrating. The brain is not an inert thing but a dynamic living organism, constantly busy doing something. Mechanical algorithms store memories of how to lift a teacup or tie a shoe lace, processing input and solving problems.

We cannot see the mind, but we can at least measure the physical properties of the brain. It weighs no more than a kilogram, possessing the consistency of cold porridge, three quarters water, the rest fat and proteins. Our smartness resides in the way these simple ingredients are organised into something infinitely subtle and sophisticated.

As it matures, nature, nurture, genes and culture help our brain to connect eighty five billion neurons, ten trillion synapses and sixty trillion atoms in one million billion different ways through a hundred thousand miles of neural wiring.

The living brain

We call the brain 'grey matter', but in reality it is red with blood, rushing oxygen, nutrients and chemical messages across synapses. Brain neurons are 'excitable', always on the lookout for an input, but ninety percent of the brain's cells encasing the neurons are 'non-excitable' glial or 'glue' cells, up to fifty networking 'secretaries' for each executive neuron. These carry out vital message-relaying roles, possibly able to communicate with each other without synapses. Einstein's brain possessed a higher than average proportion of them, so they may have a function that scientists have yet to discover.

We Are More Than Our Brains

Each neuron has twenty thousand neighbours with access to billions more through its own intranet. A frequently peddled neuromyth is that we use only a fraction of the brain's potential at any moment, but the reverse is the case: we need all of its functions all of the time, vitally inter-connected, even in sleep. The only truly silent brain is a dead one.

No matter how thorough, bottom-up descriptions of the brain's size, substance, statistics, mechanisms, parts and processes do not account for the top-down whole phenomenon that we call mind, or how we get creativity from chemistry. Neural networks operate as closed internal systems, but mind adds something beyond conventional scientific explanation, the unpredictable open experience which we call life. Viewing mind as a separate agent from its brain raises as many problems as it solves.

Metaphors of the mind

Consider for instance the various ways of staging a puppet show. In Vietnam, water puppets are operated remotely, the puppeteers invisible behind a screen, the rods that move the puppets hidden under the water. In Indonesia, the shadows of silhouette figures are cast on a screen, their body parts worked by thin sticks manipulated from below. In Europe, marionettes are controlled by strings from above, often barely concealed.

Such puppets are worked externally. This might explain why the most convincing puppets to watch are internally operated hand puppets, which seem to be motivated by a mind of their own. But even this sleight of hand does not tell us who is the organ grinder and who is the monkey. It merely leads to an eternal regress: who or what is moving the hand inside the puppet?

Puppet and puppeteer
Water, glove, string or stick puppet – who or what moves the mind that moves the puppet?

117

We Are More Than Our Brains

Our difficulty, or perhaps our triumph, is that our brain and mind are so seamlessly woven into our every moment that we never see ourselves as puppeteer or puppet, master or slave, baron or serf, with brain in the ascendant one moment, mind the next. Our mind is like nothing else in nature, so we cannot account for it other than in the language and concepts that our brain grants us.

We can refer to the kidney as a filter, because that is what it does. The brain however operates in ways unlike anything else known to us, and we get no clues to how it works merely by gazing at it. We have to resort to metaphors as poetic flourishes to help our understanding, which tend to reflect the technology of their day.

In the seventeenth century, when the industrial revolution was just beginning, the imagery of the brain reflected the world of mechanics: it is a series of pumps, valves and sluices, driven by hydraulic pressure. In the eighteenth century, more sophisticated clock-making techniques led to the creation of automata, and the belief that much of what goes on in the human body is similarly mechanical.

Darwin shifted the focus away from mechanics back to biology in the nineteenth century with his theory of evolution: it's much more fruitful to see ourselves as animals sprouting from a tree of life, not as animated machines. The modern fashion has swung back to mathematics and computing: the brain is a hard-drive, a control centre, a telephone exchange, an intelligence agency, a transmitter and receiver, a black-box flight recorder, an information processor, a combinatorial serialiser, a probability calculator, a belief engine, a neuronal potential activator and a coding vector.

Some push metaphor to the very edge of intelligibility: the brain is a quantum cohesion, at once beyond our intellect and yet the very essence of it. There is something (say the quantum theorists) in our brain that defies the laws of classical physics, a 'coherent field' that enables feats of speed and communication between cells without which plants could not photosynthesise, birds could not migrate, human brains could not think, jazz players could not 'jam', and the internet could not circle the globe.

Right reduction

Such images tempt us to see the brain as little more than bits, bytes and biochemistry. Francis Crick, co-discoverer of the DNA double helix, echoing sentiments first voiced by Hippocrates over two thousand years ago, considers it an 'astonishing hypothesis' that 'you, your joys and sorrows, your memories and your ambitions, your sense of personal identity and free will, are in fact no more than the behaviour of a vast assembly of nerve cells and their associated molecules'.

Astonishing indeed: the words 'in fact no more than' reveal the limits of over-zealous reductionism, or 'nothing buttery'. If we follow reductionist thought to its logical conclusion, we find that inside every 'associated molecule' are millions of atoms, inside which are billions of particles, inside which are even smaller 'bits' that we haven't discovered yet, all the way down. We never reach rock bottom, because there is no rock and no bottom.

Science began in Ancient Greece as atomism, or the theory that all matter, dead or alive, can be reduced to these tiny 'bits' that can't be cut up any further, randomly colliding in the void like billiard balls. After all, the only difference between water, steam and ice is how tightly their molecules are packed together. The atomists saw life and even the human mind as evolved from slime, with no need for a guiding hand or ultimate purpose.

Over two millennia later, atomism has morphed into eliminative materialism: the mind is a by-product of neural buzzing down below, in a universe made by blind forces. By over-applying such reductionist thinking to the physical world in its entirety, scientists of all stripes have assumed the mantle of secular gurus, agony aunts and soul doctors, deconstructing our values, feelings and beliefs into their component atoms, genes and algorithms. Therapy has been replaced by theory. Lived reality, with its nourishing narratives, imaginative dreams and true illusions, is laid bare for what it is: a flimsy overlay of material reality.

A glance at human history and culture suggests that this is a false, unnecessary and dehumanising ideology. There are no grounds for forcing ourselves to choose between being material or spiritual creatures, because we have evolved as both, blessed with complementary ways of knowing and being.

As the poet Emily Dickinson wrote, 'The brain is wider than the sky'. We are born with a brain, which is a matter of biology, but once we begin our life journey, experience gifts us our sense of human 'being'. We discover that, beyond being material, or flat, or value-free, 'the mind has mountains', as the poet Gerard Manley Hopkins put it.

Rightly-applied reduction has helped us to liberate ourselves from ignorance and superstition. It teaches us how wrong our forebears were to accuse the mentally ill of demon possession or bad character. Psychiatrists can now prescribe medication that frees schizophrenics from tormenting hallucinations, and depressives from overwhelming brain states.

Reduction is an essential practical tool for studying phenomena that regularly repeat themselves, taking things apart to see how they work at lower levels down the hierarchy. All disciplines of enquiry have their backstops, the 'bits' that make the 'its' of their contemplation. The physicist seeks atoms, the biologist traces genes, the logician defines necessary conditions, the banker tracks debits and credits, the moral philosopher drills down to the choosing agent.

Reduction can help us to be more astute in locating and identifying causes. Often we over-determine the causes of our actions: we're annoyed at being shown up in front of other people, or we're reacting badly to last night's curry. In the Middle Ages, William of Ockham suggested that we cut back to the minimum cause of a thing, without multiplying hypotheses unnecessarily. It turns out that we feel angry simply because we have low blood sugar levels.

Our theory-making also needs to apply reduction appropriately. Intelligent Design suggests that the world is the way it is because the Good Lord made it so, good and evil alike. But a theory that explains everything explains nothing. Why should this thing be good but that thing be evil? Darwinism by comparison is a *scientific* theory because, by being specific, not general, it gives a material explanation for how things are, accounts for differences, works from the facts, makes dependable predictions, points to new knowledge, and is open to correction.

Over-zealous reduction

A car mechanic applies reduction to find a fault in an engine, and then put it back together in better working order, but this does not mean that the engine is merely the sum total of its spark plugs and pistons. It still needs someone to drive it. Reduction can lead to useful Newtonian or Darwinian laws and theories, but we can't use it to shred our higher-level abstractions of ourselves as human persons and moral agents to operations at a level of explanation which makes sense at the bottom of a ladder of causes, but is unable to generate meaning and significance when we get to the top.

We might say for instance that the sounds of our favourite guitarist 'really exist' as electrical signals in our auditory cortex, but so does our enjoyment of the music in the act of listening to it in our chosen moment. We can't explain a higher level of reality by reducing it to its causes in a lower one.

By the same token, if we use eliminative reductionism to explain religious visions as the chemistry of proteins, hormones, neurotransmitters and blood sugars, it becomes a 'universal acid' that dissolves their significance in a personal life story, an emotional journey, an intellectual quest or a cultural tradition.

The challenge for reductionists is that knowing the position of every particle at the lowest level of explanation is no basis for predicting the future at the highest level, because we inhabit a universe of flux and energy, not fixity and stasis. We are embodied in dynamic space-time, caught in the endless dance of feed-back and feed-forward.

We can't be reduced to our genes, because although our genes change our environment, our environment also changes our genes. We can't be reduced to the electrochemical activity of our synapses, because we are participants in our own drama, which is always unfinished. Try as we may, we cannot catch our imagination in the act.

Reductionism and materialism are as much value-statements or acts of faith as any other philosophy, albeit dressed up to sound hyper-rational. As claims to theoretical truth, they face a credibility paradox: they set out to persuade us that we are not free to choose, then ask us freely to believe in them.

Atomists, materialists and reductionists overlook the plausible possibility that there are just as many reasons to trust in a more 'human' story about the evolution of the mind. Our inherent belief that we possess a self and free will, dismissed by some materialist thinkers as tricks of the brain, has served us well in the practical business of life for millennia. We might be *physically* stuck with the shape of our nose, or our poor ability at algebra, but we are *metaphysically* free to help our neighbour, or drink less alcohol. As the Holocaust survivor Viktor Frankl pointed out, in any situation, regardless of the duress we feel under, we are always free to choose our attitude.

This suggests that so-called folk psychology stands on firmer ground than materialists allow, consistently passing the tests demanded of Darwinian theory: it gives us an explanation of ourselves, and enables us to make predictions on which we can base everyday decisions.

Mythos and logos

The Greek thinkers Plato and Aristotle, the founders of western thought, rejected atomism, materialism and reductionism as total explanations. Reason gives us a version of the mind from the outside, but it has no power to illuminate it from the inside. They

121

saw order and pattern in a purposeful universe all around them, including in their own thinking, but they were also deeply conscious that there is a strong element of the irrational in human life, which it is dangerous to suppress.

They recognised that nature is not random but underpinned by a guiding principle or faculty of mind called Logos, capable of shaping what makes a logical argument, a wise decision, a just act, a thing of beauty and a rule-governed cosmos. Christian thinkers were later to rework this idea into a transcendent presence within and beyond nature, but for Plato and Aristotle it was a harmony at the heart of earthly things, not a supernatural force.

Plato exposed the poverty of reductionism by calling us a featherless biped, which makes us no different from a plucked chicken. He knew that, if we peel all the layers off an onion, we end up going hungry. Aristotle was also aware that confusing a thing's function with its form is to muddle levels of explanation.

Our modern science, deeply materialist by instinct, has rewritten the Logos in its own terms, redefining it as logic, or scientific method. Newton's and Darwin's theories act as rational binding agents, based on a core belief that matter and life are upheld by laws which are partly created by chance but mostly enforced by necessity.

So far so good, but logic cannot account for the whole of human experience. We are also creatures of strong feelings, irrational fears and deep anxieties, on full display in the dark tragedies that haunted the imagination of theatregoers in Plato and Aristotle's day. They knew that as well as Logos we need Mythos, which is the stories we invent to help us cope with life's uncertainties.

In the introduction we mentioned Popper's Three Worlds. The first world of material things is where Logos, or logical thinking, fits most comfortably. We can measure and predict them. Popper's second and third worlds however, of personal experience and human culture, are where we encounter Mythos.

Myths are cultural and social constructions, not scientific formulae, but they are no less real as narratives we live by, such as our commitment to family, our eternal wrestle with our mortality and our struggle to control our passions. Myths that address such concerns mean far more than something imagined, or 'not real'. The legends of King Arthur don't qualify as history, but even legends have their roots in the real world, in this case the determination of a nascent kingdom to fight off oppressors, and its emotional need for a national origin story.

The secret of applying reduction is knowing when to stop, and understanding why it gives different answers in each mind-world. Logos gives us brain-pleasing explanations, while mythos gives us mind-satisfying narratives. The healthy mind is able to move freely between both.

Ascending the ladder of complexity

Scientists object when they are accused of draining life of its mystery. Instead they believe that, by unweaving the rainbow, or deconstructing the nature of light, they are offering us beauty and illumination of a different kind: to go beyond superstition into the wonder of unlocking the secrets of nature, without appealing to the supernatural. They know the difference between making something simple and over-simplifying it.

In physics, reduction has led to some remarkable advances, driving the powerful digital technology we now take for granted. Stripping things back to waves and particles is not however like taking a freeze-frame of reality, making it stand still long enough for us to measure it. After analysis must come synthesis. Quantum theory, an even wackier branch of physics, shows that there is an integrated order at the heart of things, a deep connectedness between particles.

In other words, seemingly inert matter is almost alive, and as we ascend the ladder of complexity, new properties emerge up ahead that cannot be explained by reasoning backwards to isolate or account for the constituent parts. Biologists have an even harder time dealing with the dynamism of living organisms, which are as predictable as a herd of cats.

Darwinian evolutionary theory starts from a simple enough premise, based on plentiful fossil evidence: we are descended from a long line of primate ancestors, which in turn hark back to the first primitive life forms that appeared on earth. When we look at ourselves beside a chimpanzee, this claim seems highly plausible. The challenge is how far we can appeal to Darwinian evolutionary theory to 'explain' our social practices and private thoughts, 'accounting' not just for the shape of our brain but also the musings of our mind.

We are not unused to reductivism in the cultural sphere, as a form of minimalism. Painters such as Piet Mondrian, Paul Klee and Pablo Picasso have explored how we make whole meanings or 'gestalts' from minimal visual information. We can't however reverse this process, reducing the artistic sensibility to cascades of proteins, or stripping a painting down to molecules of acrylic. Our mind has not evolved to reduce interiors to exteriors, and our eye never sees less than the whole.

Darwin was too subtle a thinker to imagine that the whole can be reduced to its parts, or that to explain is tantamount to explaining away. What we see in nature is potentials and tendencies, regulated by feedback loops inside living organisms. As a result, new properties constantly emerge at every level of organisation. Life points towards complexity, not simplicity, towards interrelatedness, not separateness, towards mutuality, not self-interest.

Nevertheless there was a fashion for a while among some evolutionary theorists of believing that reductionism can lay bare not just the workings of nature, but also of the human soul, as if total knowledge of initial conditions and prime causes can enable us to predict everything that will happen tomorrow, or vice versa. There was a mistaken belief that to understand one small detail of human genetics is the key to understanding the whole of human nature.

But knowing what Beethoven ate the night before he composed his fifth symphony will not reveal to us its majesty, or explain its emotional impact upon us. Once we've identified all the chemicals in our brain, located all its particles, decoded all its genes and mapped all its pathways, we might instead come to our senses and acknowledge the existential gap between analysing and experiencing.

Organisms are so-called not just because they are biologically internally organised. They are also active participants in a complex web of relations with the life forms around them. Technology may enable us to fertilise an egg in a test tube, but it remains the case that the womb is the best environment to nurture it, and the family is the best way to raise it.

The mind is a computer
Beyond physics, chemistry and biology, Silicon Valley has offered us a computational model of mind, on the basis that pre-wiring gives us insights into how our mind works. In the last few decades, we have come to depend on algorithmic dependability for safe banking, fast information and happy messaging.

In the Age of Artificial Intelligence or machine learning, the brain is presented to us as a pot-pourri of innate learning programs, or algorithms perfected over millions of years of evolution, propelling us from stupidity to smartness with little effort required of us. Our brain does most of our thinking for us, allowing us to go through a whole day without having an original thought.

Learning Algorithms in the Brain

SENSEMAKER **Meanings** **Causes**	Convert sense data into knowledge, use contextual clues, discriminate between appearance and reality, apply reason to experience, attribute causes and identify effects. *Is it a bird? Is it a plane? What was it last time? How do I know that a stick that looks crooked when half-submerged in water is really straight?*
MATCHMAKER **Likenesses** **Theories**	Look for similarities, patterns, family resemblances, associations, correspondences, analogies, categories, laws. *If an apple is 'pulled' to the earth, is the moon pulled in a similar way? Can we formulate a general law of gravity from this?*
HYPOTHESISER **Predictions** **Beliefs**	Anticipate, plan ahead, speculate, infer, generate possibilities and probabilities, use the past to solve the problem of the future, intuit what others might be thinking. *If we pay benefits to the unemployed, are we encouraging a dependency culture? If we don't pay, are we creating a poverty time-bomb?*
ANALYSER **Explanations** **Systems**	Dismantle and reassemble, put into order, restructure, prioritise, process information, rationalise, calculate, decode, serialise, generate a big picture from small detail. *What are the causes of road deaths? How can they be reduced? Should we focus on driver training, traffic calming, speeding fines or safer vehicles?*
SYMBOLISER **Concepts** **Ideas**	Turn impressions into thoughts, make a mindscape, think metaphorically, interpret creatively, make imaginative leaps. *This hammer with its weight and force can be an extension of my arm, a tool to shape the world, an emblem of power, an instrument of oppression.*
STORYTELLER **Connections** **Myths**	Link cause and effect in a narrative, sequence events, experiment with time, improvise endings, ascribe motives, project outcomes, invent true illusions in alternative worlds. *How did I end up here doing this? Is there an alternative ending? What if I hadn't married Sara? What if we divorced each other?*
SYNTHESISER **Values** **Wholes**	Fill in gaps, combine parts into wholes, network what is known, turn facts into values, make judgments, arrive at a unified theory, understand. *What can I believe? What is my life's work? Is there a Grand Theory of Everything?*

We can do this because evolution has bequeathed us two secret weapons. The Baldwin Effect ensures that clever tricks learned by our ancestors are added to our genetic programming. Bayesian theory, which is really sublimated mathematical calculation, gives us the confidence to throw a peanut in the air and catch it in our mouth in front of our friends.

These learning algorithms make life so effortless that some social psychologists describe the mind as being 'flat'. Once we've learned the basics, activities such as walking, talking, attributing cause and effect, predicting motion and reading other people's minds are written so deep in our genome, operating so far below our radar, that they require at most five percent of conscious input, until we are pushed outside our comfort zone.

Unless we consciously intervene, our brain takes charge of our day, as the invisible agent behind most of our actions and assumptions. It cleverly persuades us that the world has been made to fit the way we perceive it, because having made its mind up how things work, it is too lazy to start over for any reason. This means that 'we' get lazy too, too cosy in our complacency to question whether we see things as they really are.

Lazy as our thinking may be, AI engineers trying to design self-driving cars have realised that our cognition is anything but flat and particulate. The challenge they have to solve, after they have reverse-engineered all the brain's learning algorithms perfected over aeons, is to find the Master Algorithm that holds everything together. What makes our learning 'deep' and joined-up?

Consider how many programs have to be open to read 'Lfie biegns at fotry', as well as to appreciate the cultural sentiment behind it. Our brain does this without thinking, or to put it slightly differently, we don't notice its effortless unity until we hit a problem, or something that doesn't fit.

Human thought achieves its integration through parallel processing, or monitoring activity across many neural networks simultaneously. Only a small portion is carried out as linear, digital and algorithmic programming, so a computer model of the mind is not a helpful one. Unlike a computer, the mind is a complex system capable of self-organisation, like a storm, whirlpool, laser beam, friendship or immune response.

Such systems are endowed with 'attractors' that give rise to new qualities on a supra-organisational level, and this 'more' is

126

quintessentially different from its components, because it expands in a non-linear way. These features disappear when we name the separate elements, just as a dance vanishes when we write out the steps. Similarly the wholeness and depth of the human mind vanish when we reduce them to algorithms.

And yet, as the 'Learning Algorithms in the Brain' table demonstrates, our mind contains algorithms aplenty, like so many blades tucked neatly into a multi-purpose penknife, as we shall consider later. These do not necessarily form a hierarchy, but notice the levels of omnipotence, integration, intuition and imagination involved in the final three.

Some AI afficionados assure us that we are approaching the Singularity, or the moment when possibilities become infinite. Artificial minds, capable of self-organising and out-thinking us, will appear first in our pockets, and then inside the flesh of our brain. Who knows, we might evolve into 'thinking heads', with no need for a body at all.

To this end, our brain has been repackaged to a cybernetic or business model. It is staffed by monitors, supervisors, executives, representatives, agents and managers, all chasing their targets, tailoring their economies and waving their flow diagrams.

These flunkeys, we are assured, will be not just exponentially more intelligent than us, but also differently smart. Through machine learning they will go beyond what we initially program them with, until they know us better than we know ourselves.

It remains a moot point whether our machines will instal in our minds algorithms that help us cope better with depression, international strife, environmental degradation and boredom when they put us out of work. Perhaps then will be the time to remember that there is an older, poetic narrative of how our brain works: it is an enchanted loom, a theatre, a news room, a private cinema, a council chamber, a court of law, a parliament, a shopping mall, shoots and tendrils, a Jackson Pollock painting of squiggles endlessly interconnected. Our neural networks are fountains, constellations, libraries and art galleries.

Murdering to dissect
So long as we remember that metaphors such as these can only offer us a picture of the brain *as if*, they are helpful. The problems begin when we start to take the metaphors literally, or see them as reducing instruments. This is because, as far as the coding of the brain is concerned, a metaphor is a descriptive analogy, not an explanatory mechanism of any kind.

We Are More Than Our Brains

In the brain, '$E = mc^2$' is encrypted in the same bytes as 'I love you'. What to the mind is vibrant colour is to the brain only frequencies of light waves or vibrations at forty Hertz. The eye senses polychrome inputs and outlines, but the mind enjoys the epiphany of standing in front of a Leonardo painting.

Dissecting the acorn destroys the oak, because no single brain molecule carries a mathematical insight, moment of desire or the secret of life, just as no single blade of grass makes a lawn. A lawn is a human artefact, attentively mown and watered, responsive to the feet that stroll and play upon it. If we muddle levels of explanation, or assume that the complex masks something very simple, we have no way of discriminating between the smile of the Mona Lisa and the striped wallpaper in our living room.

The reductionist whittles the aesthetic sense down to pattern detection, but Leonardo portrayed beauty as the proportion of Vitruvian Man, Charles Darwin attributed it to the healthy forms produced by natural selection, and Werner Heisenberg saw elegance in the design of the cosmos. Sigmund Freud had the wit to remark that, though beauty has no obvious use, civilisation would be unthinkable without it.

Beauty is an emergent feature of the world, a creation of the mind, so there is little point looking for the neural hard-wiring or genetic coding that is 'responsible' for it. Similarly, there is no mechanism of one-neuron-one-spark or one-gene-one-enzyme for making us develop into a psychopathic killer, or believe in God. Even if we find the 'God spot' and stimulate it with a probe, we are a million miles away from understanding what spiritual experience *means* or *feels like*, because this can happen only subjectively, within a particular culture and tradition.

When pushed to the edge of its capacities, or close to oblivion, our brain might implode and give us a near-death or 'out of body' experience, resulting from a malfunction in the angular gyrus, but whatever the bio-chemistry of the moment, and wherever in the brain it happens, it is still an *experience.*

There are no explanations, only ways in which we mix the fine ingredients of our perceptions with our interpretations, and the choices we make based on them. 'The brain gives us our reality', intones the neuroscientist, to which the philosopher replies, 'Our mind turns it into a habitable world'. Somehow, from billions of separate brain events, each insignificant if isolated from its neighbours, the phenomena of intelligence, consciousness, personality, spirituality and culture assume meaning in the context of a human life.

How does our brain manage its economy?

Energy – electricity - the nervous system – emotions - the immune system – cancer - antibiotics

- Electricity gives our nervous system its speed of response.
- A knee-jerk reaction is a one-way reflex arc, but our motor system is a two-way voluntary response: we don't have to put food in our mouth every time we see it.
- The spine is the central highway of our nervous system, and paralysis results if traffic is blocked in any way.
- Our autonomic nervous system operates our internal organs automatically below the radar, though yogis can learn to regulate their heart beat by the power of thought.
- Emotions are default settings, but feelings are our personal responses to incoming messages: we don't *have to* get angry about that nuisance phone call.
- Our body has co-evolved with millions of bacteria and viruses.
- Our immune system is primed to attack germs that pose a threat to us, but sometimes it over-reacts, destroying friendly cells.
- Early exposure to pathogens gives us stronger resistance to infection as an adult, and possibly better protection against cancer.
- It pays to keep our gut flora healthy, be sparing with antibiotics, and ensure that our brain is well oxygenated.

Powering the brain
A brain the size of ours in proportion to our body weight is a hungry beast. In adulthood it demands fifteen per cent of the body's oxygen intake and consumes twenty per cent of its metabolic energy. A child's brain consumes forty percent, because evolution has prioritised brain-building over physical growth in the early years. Human childhood is lengthy, not because our body

is slow to grow, but because, through a process called neoteny, we stay younger for longer, so that our learning potential can be extended indefinitely. We need more sleep in childhood because our brain commandeers up to sixty per cent of the body's genes to construct itself, gobbling up as much as half of our basal metabolic rate.

When fully grown, our brain demands forty per cent of our blood glucose as a constant supply, because it can't store it. A quart of hormone-rich blood must flow through our brain at any moment, replete not just with sugar and oxygen but also hormones.

There is no point at which the brain idles. Even in sleep it is a whirl of activity, sorting, dreaming and generating some of our most creative ideas. On average it consumes twenty times more fuel than a muscle at rest, humming along at forty cycles per second, burning between two and four hundred calories a day, more if serious thinking is demanded of it.

Given its insatiable appetite, the big brain must have evolved for a big reason. In the natural world, energy-rich brains are an important source of protein. When chimpanzees catch a colobus monkey they eat the brain first, and for all we know some of our ancestors cannibalised each other for the same reason. Eating brains does not alas raise our IQ, only our energy levels.

The wet and the dry

There are as many as six trillion chemical reactions going on in our body at any second, most of them masterminded by the brain through the 'wet' delivery system of the blood stream, ferrying action-triggering hormones around the network and across synapses. This is a relatively slow process, as we discover when we pop a pill and have to wait for it to 'kick in'.

Most treatment for depression is 'wet', and involves getting the cocktail of drugs right, which might take weeks. The advantage of 'wet' chemicals over 'dry' electricity is that they are more flexible, delivering not just on/off signals to the synapses but also allowing time for the mind to become aware of subtle changes in consciousness.

'Dry' electricity fires the machine much more rapidly, both human and robot. We are made aware of this through the hypnic jerk, when our whole body jumps involuntarily in bed, usually when transitioning into sleep, or perhaps linked to a moment of 'falling' in a dream. Amputees can be fitted with artificial limbs that detect the tiny electrical charges given off by the brain, and translate them into corresponding movement. Technicians are perfecting machines with sensors that can read our minds, sufficient at least to power prosthetic limbs.

Electricity is the biological battery that amplifies the power of the brain, the material base of what the ancients regarded mystically as the spirit that moves us, in this case quite literally. There are as many as five different types of electrical transmission in the body. Its advantage is speed: electric pulses give the brain its rapid recognition of intruders, instant response and quickness of thought, though the super-quick electrons in our brain cells are only the carriers of our thoughts, not the authors of them.

The role of electricity in powering the brain was demonstrated dramatically by the physiologist José Delgado in 1963. He faced a rampaging bull in the ring, but he had earlier fitted what he called a 'stimoceiver' in the bull's brain. In his hand he held a transmitter, and by sending a radio signal, he was able to stop the bull dead in its tracks.

This was a bit of showmanship, and planting electrodes in the brain is invasive, with potentially sinister implications for thought and mind control. Sufferers of schizophrenia and Parkinson's disease however have shown in experiments that the ability to self-stimulate can help them regulate their symptoms.

At the synapse, or the gateway to every neuron in our brain, the transmission of signals is technically neither wet nor dry, but electrochemical, which is a mixture of both. This is achieved by ions or molecules that pick up a positive or negative charge from their neighbours, triggering a response as they move back and forth across the cell membrane. This is happening millions of times a second in our brain, though we are never aware of it as felt experience, except when a critical mass is reached. Then we have to decide whether to buy that burger, or stick with the diet.

The nervous system
Electricity powers the current surging along the axons or connecting threads of our nervous system, carrying electrical charges at a steady rate back to the brain. If all the axons in our body were stretched out in a continuous line, they would reach to the moon and back. The axons in our brain are very short, and not every neuron is connected to every other, otherwise the brain would be overcrowded. The longest neurons in our body are found in our spinal cord, their axons running all the way down to our big toe, though these are dwarfed by the length of the axons connecting a giraffe's brain to its tail.

Feel the force
Speed of response, pins and needles, reflex reactions, jumping in our sleep, sharp pains – all remind us of the electrical sparks that fire our body.

Regardless of length, axons get their messages through, because they work on an 'all or nothing' basis, either 'on' or 'off'. They are like trails of gunpowder: once lit, the message is going to cause an explosion at the other end. The explosion doesn't have to be violent: it may be an ant tickling our toe. If we drop a brick on our foot, it will cause a much bigger bang in our brain, triggering sharp pain.

Signals don't get lost, whatever their strength, because they are boosted along the length of the axon in a relay mechanism, or frequency of firing. Occasionally the whole system can misfire, as when we suddenly 'jump' in bed, or suffer an epileptic attack.

The only way the signal starts to feel weaker is through habituation: the brain ceases to notice the message. If not, pain would never wear off, and we would be constantly aware of our clothes rubbing on our skin. Some stimuli are too important to be allowed to wear off. Our body might gradually adjust to the cold water of the swimming pool, but a hand thrust into the fire will produce nothing less than screaming pain until it is removed.

Repeated stimuli are essential to how we learn. Not only do they increase 'potentiation' of the synapses, they also cause our super-conducting axons to thicken and develop white myelin sheaths, forming en masse the 'white matter' of our brain. This insulation prevents short-circuiting and speeds message flow a hundred fold to create 'learning super-highways'. This process can occasionally go wrong. Multiple sclerosis is a disease of the autoimmune system when the body attacks the essential myelin sheathing of its nervous system, resulting in gradual difficulty with movement.

We often talk of our grey matter, comprising our neurons and synapses. It is our white matter that is the true secret of our smartness, made up of all the axons that bind our brain together, spreading through it like the fine filaments of fungal spores.

Our brain starts to form as a tiny neural tube three weeks after conception, maturing in adulthood to a three pound corrugated mass at the apex of our nervous system, all roads leading into and out of it. It absorbs six times more in-coming information through its sensory system than it emits as out-going motor messages, whether as a raised arm or spoken word.

Sensory feedback travels along the same nerve channels as commands to move our muscles, but when it gets to the brain, it forks to different processing areas. When we pick up an unexpectedly hot dish, our brain freaks out. We want to put the dish down as quickly as possible, because it hurts, but we don't want to drop it, because it contains our dinner.

Pioneering neurology

It took many centuries and some pioneering neurology by Lord Adrian in 1928 to understand the connection between the outside world and the brain, linking the interplay of neurons, synapses and electrical signals. He established that the brain is the prime mover of our whole body, because all signals begin and end there. Some messages are one-way, like the reflex jerk of our lower leg when our knee is tapped. Most are two-way, such as when we quickly swat that wasp that has landed on our arm.

Adrian showed that a nervous message looks nothing like its stimulus, because it is made up of coded electrical pulses that are decoded by the brain. The strength of signal, determining how much it hurts, is decided not by the size of the spike, but by frequency of firing. Cracking our skull on the door frame sends far more signals to the pain centres in our brain than a fly landing on our scalp.

When we squint in bright sunlight, it is because more light information than usual is streaming into our retina and being passed as electrical signals to our visual cortex. We can prove that the signals are more important than the incoming light by a simple experiment: when we press on our eyelids, we can induce ourselves to 'see stars', even with our eyes closed.

In many ways our nervous system is a brain in itself, as intelligent as its boss, capable of handling a huge number of variables, far more harmonious than any human society. A creature with a simple nerve net such as a starfish can get by with a tiny control centre located in its stomach, but not so a human.

A hydra, a very simple aquatic animal barely a centimetre in length, has no single brain in one place, but three hundred neurons spread through its whole body acting as a nerve net, deciding whether it should retreat from a threat or advance towards a potential food source. Scientists believe that if they can map how this proto-brain controls the animal's movements, they might find the key to unlocking the infinitely more complex human neurome.

This would be a mighty achievement. It took decades to decode the human genome, but mapping our neurome may well take a century. A big brain like ours can't function without a highly sophisticated nervous system, and vice versa. Imagine the degree of co-ordination in a ballet dancer as she glides across the stage. She is far more than a bundle of reflexes, more a display of brain-and-body synchrony as she effortlessly toggles between pirouetting on her own body weight, then becoming weightless in the arms of her partner.

There is an obvious virtue in having a strong command-and-control centre in one place with a 'hot line' to all areas, though it makes the whole body vulnerable when it receives a head injury. An alternative model is the octopus, belonging to a completely different branch on the evolutionary tree from us. It has eight brains, one in each tentacle. In this sense, its brain is distributed right through its body, which means one of its tentacles can continue to strangle us even after we have severed it. A chicken can continue to run around the yard after it has been decapitated.

Getting messages in and out
Our central nervous system is precisely what its name implies, a main highway of information running up the spine to our brain. Incoming messages from our body's sensors are called afferent. These are by no means all external, seen only with our eyes or felt on our skin. Some come from deeper inside, such as hunger pangs or butterflies in the stomach.

Some messages come from our muscles, giving us proprioception, or our sense of where our limbs are in space. When we reach out for an object in front of us, especially in the dark, our brain needs feedback so that it can 'see' on its body-map where our fingers are. When we pick up a baby, shake someone's hand or grab a racket to play tennis, our brain needs information from our fingers so that it can calibrate how gently to caress, how firmly to squeeze, or how tightly to grip.

Without afferent or incoming sensory information about the outside world, our brain could not form perceptions or make thoughts. Having done so, it needs to reciprocate by generating outgoing or efferent responses to activate our muscles, and this calls for specialised cells in our nervous system called

134

interneurons, running the length of our spine, connecting mind and body, converting incoming signals to outgoing messages.

Efferent messages complete the circuit by carrying instructions to the motor system of our muscles: we raise our arm to put some food in our mouth, or steady ourselves. Poliomyelitis is a disease spread by a highly contagious virus that destroys the delicate motor nerves that carry the brain's messages to the muscles. For generations it crippled many children, but an intense global vaccination programme means it is now limited to three countries.

Not all our responses are voluntary. Reflex arcs generate responses beyond conscious control, such as blinking, or ducking an object coming fast towards our head. There are also conditioned reflexes, such as feeling sick when we smell curry. This will probably go back to that buried memory of the time we ordered a bad curry, and felt violently sick afterwards. Or we might always perk up when we hear curry is on the menu, because it reminds us of our childhood, or so many nights-in enjoying a takeaway with friends.

Mediating feelings

Such occasions remind us that we are not simple input-output processors controlled by the strings of our nerves. Normally, between the input and the output, something important happens: we work out our *feelings* about the stimuli in our surroundings. Do we want to advance or retreat, are we excited or afraid? This marks the essential difference between a machine, an animal and a human being. A machine responds automatically, an animal responds instinctively, but a human being responds consciously and personally.

Emotions are blanket responses, but we personalise them as feelings about what is going on around us. Anger is a general emotional state, which we can trigger just by clenching our fist or gritting our teeth. It is a different question how we control, justify or express our wrath or irritation. We might lose our temper, hide our indignation, forgive the transgressor, or take calm steps to ensure that whatever caused our anger doesn't happen again.

This mediation of our perceptions by our mind frees us from being creatures of instinct. The philosopher Jean Jacques Rousseau credited us with a quality he called 'perfectibility', or the capacity to impose our will on nature, not merely reacting like Pavlov's dogs.

This potential for delay and reflection is essentially a kind of thought process, giving us a degree of conscious control over our responses, making us less likely to be hijacked by our emotions.

135

This is why we encourage children to verbalise their feelings, hard though that is. Even to say 'I am upset because she took my toy' creates a calming distance between a provocation and our response to it.

Emotions emerge spontaneously from our viscera, making them hard to stop or hide once they have kicked in. Airport security cameras can be programmed to detect unconscious facial 'micro-expressions' which might give away fugitives, criminals or terrorists.

Sometimes our emotions leave us feeling powerless, one overriding another: embarrassment can quash our sex drive, disgust can kill our appetite and shame can extinguish our wrath. With practice however, we can gain some control over these subterranean rumblings. When we admire a painting of a nude figure, we can feel moved by its form and beauty without feeling sexually aroused by its nakedness.

Our emotions have evolved to kick in a few milliseconds before our thoughts, just in case there is a threat. Without emotion, every experience would feel the same, pain and pleasure would be indistinguishable, and we would not be able to decide our attitude towards a stimulus. Life would be colourless, and we would be unable to make even simple decisions.

Our emotions come first in nearly every sense that matters. When we see a dog, it's never just a dog: it's my dog, a friendly dog, a stray that might bite me. This is important for understanding our perception: what we see is always accompanied by, perhaps even preceded by, our attitude towards it.

Even the words we speak are coloured by the feelings they evoke. Tears and emotions evolved before language, which is why we find it so hard to put our feelings into words. This means that our reactions to others and sensitivity to their feelings emerged early in our evolution. Emotions also come first in the mother-child dyad. Close emotional interaction is essential in the early months for later neurological development and sociocultural growth. We've even had to invent emojis to add emotional tone to our emails.

Humanising emotions
Emotion is a universal human language, as Darwin originally suspected. Since his time, psychologists have agreed on six basic emotions, consistently recognised in facial expressions across all cultures: anger, fear, surprise, happiness, sadness and disgust. These are biological imperatives. Some researchers add joy,

136

shame, jealousy and guilt to this list, though these tend to show more cultural variation.

Disgust is often portrayed as a 'lower' emotion, but as sensitivity to 'bad taste' it played a vital evolutionary role in making us move away quickly from harmful foods, dangerous parasites and potential pathogens. Usually disgust is short-lived, returning to normal after we move away from the offending article, but it can become its own toxin if it embeds itself as our chronic or default response to other human beings. In India, higher castes view the Untouchables as distasteful and polluting, and yet it is these very people who have to overcome their disgust every day to clean the toilet bowls of their so-called superiors.

'Higher' emotions can be cultivated to become longer lasting traits such as hope, resilience, kindness, optimism, gratitude and cheerfulness, qualities which the 'positive psychology' movement seeks to encourage in all of us, a far better response to life than taking anti-depressant pills. Negative emotions such as suspicion, fear, frustration, anxiety, guilt and sadness all too easily prevail if we allow them to do so. They are also easily preyed upon by demagogues who offer scapegoats or easy targets for our unhappiness, failure and anger.

We warm to spontaneity, sincerity and emotional honesty, and it used to be believed that venting our anger or 'letting off steam' was a way of reducing it, or getting it out of the system. Psychologists now agree that the opposite is true. Brain scans and blood tests show that giving way to an emotional outburst makes us more likely to 'lose it' next time, so as Aristotle advised, it is wise to keep our emotions in check. Not for nothing do our parents try to offer us models of emotional calmness when we are young, and advise us not to make key decisions when we are emotionally upset.

Emotion is often portrayed as in a constant tussle with reason, but they work in combination, not in competition. Both involve cognitive processing, but give us different motivations for our actions. This means that it is not helpful to think of one as 'higher' than the other. We can refine our feelings if we wish, and some of our reasoning can occasionally be base.

There are no 'pure' or 'raw' emotions such as anger or love, only those informed by our reasons in the context of a busy life. Even when we are at our most 'het up', we are still performing a kind of reasoning. This explains why we can be passionate about something one minute, then quickly yield to cool detachment the next.

We Are More Than Our Brains

We can make what we think is a calculated decision to give to a charity, vote in an election or be critical of a friend, but in fact we're expressing emotional attitudes about goodness, truth and justice that we have already processed at a deeper level. Our 'heart' sends us a signal before our 'head' has time to verbalise it, so the feeling of acting reasonably might be just a trick of neural timing.

More than anything, our emotions allow us to make choices. Experiments with rats show that if their peripheral nerves are deactivated, but their central nervous system left intact, they can still move, but their learning and survival chances are greatly impaired because they have no way of preferring one course of action to another. They simply don't care, because the feelings aren't getting through.

Ironically this severance of body from brain has the reverse effect for sufferers of locked-in syndrome. They are conscious of their situation, but suffer less emotional turmoil because their emotional feedback loops are disconnected. This means they come as close as possible to being a pure mind, and yet are unable to express it.

Below the radar
The peripheral nerves are the outposts of our body, feeding into our spinal cord and up to the brain, so that it can send instructions back along the motor muscles. Our somatic or body-related awareness is voluntary, allowing us to feel our every move. Without it we could not lay claim to any kind of free will. We persuade ourselves that, when we move our arm, we are telling it to do so, not our brain.

Equally important is our autonomic nervous system, which is involuntary and below the radar, connecting our brain with its internal organs. If we are punched in the solar plexus, we automatically suffer a decreased supply of blood to the brain. Whereas our motor nerves excite us to action, our autonomic vagus or 'wandering' nerve is a calming influence, connecting the kidneys, spleen, liver and heart, directing hormones to where they are needed, ensuring that infections are tackled, reducing inflammation, regulating our breathing, sleep patterns and bowel movements. It connects our head to our heart, telling us when we are tired or hungry.

We can't consciously control our kidney, liver or immune system, or feel our insides going about their business. Our brain takes care of these things automatically, though there are exceptions. Australian Aborigines can maintain their body temperature on a freezing night in the desert by going into an intense meditative state, but the rest of us have to put on extra layers of clothes.

The new science of neurofeedback is discovering that the mind is capable of altering its brain state, such as when we whistle a happy tune to cheer ourselves up. Feedback can also feed forward, research showing that merely *thinking* we are stronger can increase our muscle power by up to eight per cent. Massage and reflexology remind us of the intimate connections between our physical well-being and mental state. Our body-mind is a commonwealth, and it doesn't pay to favour one part at the expense of another.

Some believe that if we are better informed about how our brain works, we can take control of our pain tolerance levels, like fakirs walking over hot coals. By improving communication between our unconscious hypothalamus and conscious neocortex, we can ramp up our mind power or use neurofeedback to overcome 'psychosomatic' symptoms such as back pain, allergy, addiction, depression and phobia.

Spinal injury

But this common-sense view might be naive. When we injure our spinal cord, we are reminded that our voluntary movement is dependent on unbroken links between our brain and its motor nerves. Afferent or sensory nerves carry messages towards our brain, but in order to move, we need efferent or motor nerves to convey signals back to our muscles. If our back is broken, this two-way flow is seriously compromised, causing loss of movement and feeling.

Everything depends on how high up the damage occurs on the spine's ladder of thirty one pairs of nerves. A mid-spine injury leaves us the use of our arms, but not our legs. A high-spine injury, such as happened to 'Superman' actor Christopher Reeve through a riding accident, causes the loss of movement in all four limbs.

He did not die, because his twelve pairs of cranial nerves at the very top of his spine, their axons entering straight into his brain above the spinal injury, continued to feed his eyes and ears information about the world, and his vagus nerve still regulated his autonomic system. He could still live a full mental life, because although his somatic system was broken, key parts of his central nervous system were still active.

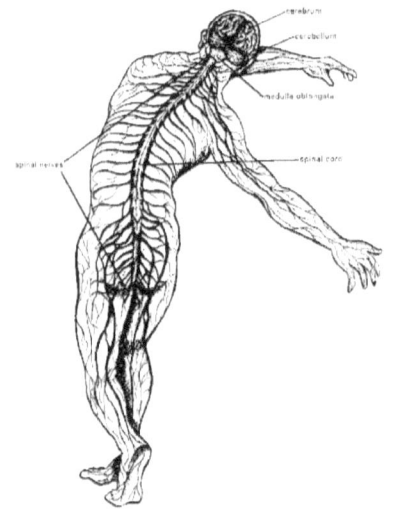

The highway of the body
Nerves enter the spine all along its length, and then up to the brain. Paralysis after injury to the spinal column depends on how high up the break is.

Quadriplegics with full spinal cord injuries have motor difficulties performing sex, but they can still feel desire and experience orgasm as hormones are carried to the brain through the bloodstream. The vagus nerve offers an alternative pathway from the genitals to the brain, bypassing the shattered spinal cord.

Every case of paralysis is different, depending on the extent and location of the damage, and how many messages can get through. Some might be able to feel their limbs but not move them, others vice versa, and levels of pain are similarly affected.

It has long been a medical dream to get the lame to walk again, and there are various avenues of research. Tiny electrical implants might be able to restore the severed links between the brain and its body. Virtual reality has had some success in improving mobility and feeling in paralysed limbs by retraining the brain to regain lost connections. Nerve repairs have been made from synthesised spider web, the finest and most flexible material known in nature.

Paralysis might kill 'action potential' in the motor system, but not emotion, because the body has other ways of communicating with its brain. The bloodstream continues to ferry 'wet' chemical messages back and forth, so Christopher Reeve could still enjoy bodily talk-back through his electrochemical systems. He could still feel joy and sorrow, so his world was not emotionally flat.

We Are More Than Our Brains

Our body defences

Our immune system is our body's natural defence against invaders, primed over millions of years of evolutionary cat-and-mouse games with billions of bacteria and viruses that outnumber the cells in our body. Two thousand species of bacteria inhabit our gut, and that's just those that have been identified. One of the breakthroughs in our evolution was bacteria learning to cooperate with the mitochondria in our cells to form alliances that benefit both, such as producing all the proteins to make the neurotransmitters that run our brain, but others can kill us outright if they break into the wrong part of the system.

Our immune system works in cahoots with our endocrine system, which controls the flow of life-maintaining hormones and protective antibodies around the system. Together they act as a kind of brain within the brain, or pre-brain, in constant crosstalk, aimed at sustaining the fragile cosmos of our body amidst the ever-threatening chaos that surrounds it.

The brain is often compared to a machine or computer, running on 'dry' electricity. At the synapses however, things get 'wet'. Our brain is more like a gland than a machine, operating a mixed wet/dry economy, its messengers negotiating dry tracks of electrical nerve impulses one millisecond, and fording wet streams of biochemical activity at the synapses the next.

Some brain researchers concentrate on studying the fast dry nervous system, while others focus on what happens in the slower moist spaces between the synapses. Electricity shooting through the nerves outpaces the sluggish blood stream, but together they make a rapid response system, messages passing very quickly between neurons and into the body, as we discover when we get an adrenaline rush.

'We' have nothing to do with the efficiency of our endocrine system or the xenophobia of our immune system. They are controlled centrally by the master pituitary gland, regulating the release of around two hundred complex chemical compounds or hormones into our blood stream, their impact on our brain and body so powerful that some of their names are part of our everyday vocabulary: oestrogen, testosterone, dopamine, oxytocin, serotonin and adrenaline. Twenty of these cocktails are produced in the brain, others in the thyroid, thymus, pancreas, ovaries or testes, designed to regulate our appetite, temperature, sexual urges, moods and stress levels.

Our immune system complements this chemical flow by fighting a constant war against intruders. The linings of our nose contain millions of exosomes or invader-spotting cells which

141

secrete sacs of fluid to repel marauders before they get on board. If insurgents get past this first defence, they face a host of white blood cells armed with antibodies primed to zap them.

As well as conflict, there is also symbiosis. Over millions of years our body has fought a war with pathogens such as viruses, bacteria and intestinal worms. Sometimes we lose a battle. We have no defence against a parasite found in cat litter that wriggles its way into our brain and makes us prone to excessive risk-taking. Mostly we turn offence into defence, as in the case of the milkmaid who didn't catch smallpox because she had already been exposed to cowpox. We live in fear of germs, but in reality they have forced our body to be as fit as it can be, and incited our brain to perform to ever higher levels.

As a result of our long-running war with parasites of all kinds, our body hosts more bacteria than its thirty trillion cells. Millions remain on our hands even after we have washed them, and we use 'friendly' ones to create the flavours of our bread, cheese and beer. Once inside us, the vast majority form a confederacy with our body cells, or live inside them, helping us digest our food and metabolise, any aliens challenged by sentries on constant alert.

A bacterial or viral infection results when this vigilance fails, and can prove fatal. It's not so much the invaders that kill as the toxins they produce in the body. The bugs we hear most about, and do our best to avoid, are those that cause cholera, tuberculosis, tetanus, pneumonia, meningitis, sexually transmitted diseases such as syphilis and gonorrhoea, and food poisoning such as listeria and salmonella.

A bacterium is a single-celled organism which tries to overwhelm the healthy cells in our body, and can be killed by the right antibiotic. These do not work with a virus, which is a nano-strip of RNA designed to unlock the door of a cell's protective coating. It is 'alive' only when it wheedles its way in.

Like bacteria, we have befriended some viruses in our long evolution, making up about ten percent of our genome, so they too have helped to drive human evolution forward, keeping our heart healthy, fighting off cancer and protecting against malaria.

Sometimes we come second in our fight against parasites. Retroviruses, buried in our ancient DNA and usually switched to 'off', can be reactivated during times of stress, and have been linked to diseases such as multiple sclerosis and schizophrenia. The chicken pox virus can lie dormant in the spinal column from childhood, then be unexpectedly triggered as shingles years later.

We Are More Than Our Brains

A deadly war

The viruses we have to be particularly wary of are those that jump species or dangerously mutate, such as SARS, MERS, Ebola, Zika and other forms of Coronavirus. Animals suffer from influenza too, and so far we have had outbreaks of chicken, bird and swine flu in human populations. It is believed that the AIDS virus originated in chimpanzees.

Viruses from other species are unknown to our immune system, leaving us defenceless and vulnerable to infection. The Black Death in the Middle Ages, which exterminated over half of the population of Europe in a series of outbreaks, was probably caused by a bacterium transmitted by rats and their fleas. The so-called 'Spanish flu' pandemic of 1918, which killed an estimated fifty million people, many of them young, was probably caused by a bird flu virus.

The viral infections of smallpox and polio were as good as eliminated in the twentieth century, but almost exactly a century after Spanish flu, Covid 19, jumping from species such as bats or pangolins, held the world to ransom. The race began immediately for an effective antiviral vaccine, but it takes months if not years to develop and test, made more problematic because viruses, like cancer cells, are clever at disguising themselves and mutating.

At any one moment we might be carrying up to a dozen nasty viruses, any one of them capable of killing us. We never find out, either because our defences keep them ring-fenced, or we feel only slightly ill. It is not in a virus's evolutionary interest to kill us too quickly. We might be infectious but show no symptoms. Asymptomatic carriers are a virus's best friend, and a nightmare for health experts trying to contain an outbreak.

In the last thirty years the Anti-Vax movement has persuaded many mothers not to vaccinate their children against measles on the grounds that the vaccine is linked to the onset of autism, a claim that has since been exposed as untrue, if not deliberately fabricated. Vaccination primes the immune system to fight a specific interloper, and leaves an 'immune memory' that can give a lifetime's protection.

Some early vaccines did have dangerous side effects, and it is normal to feel slightly unwell after a jab, but campaigns against safe and effective modern vaccines that have saved the lives of millions are based on pseudo-science. As a result, many who were not vaccinated as children are contracting a more severe form of measles in their adulthood which sets back their immune system by up to ten years. Anti-Vaxers insist that we cannot be forced to be healthy against our will, but they fail to think long term, and in the wider public interest.

Viewed from a Darwinian perspective, a 'superbug' is bound to kill those with the lowest defences in a population. The old are especially vulnerable, because the immune system weakens with age. This leaves only the strong to pass on their resistance, until eventually there is 'herd immunity'. This can take several generations, inflicting much suffering and many casualties in the meantime, a prospect which no modern society with the technical and medical expertise to fight the disease can contemplate.

The glands of our auto-immune system detect threat from both outside and inside. Outside, excess sunlight on our skin provokes an immune reaction to prevent burning through the production of melanin pigment, which we more fashionably call a suntan.

Inside, protecting the body involves secreting antibodies to attack aliens, especially those coming from other people. After an organ transplant or skin graft, there is a higher chance of success if the recipient is first fed a cocktail of the donor's microbiotic bacteria, giving their body time to adjust to what it would otherwise perceive as lethal invaders, and attack with antibodies.

Some intruders we can't defend against, such as snake venom, but our life can be saved if we are injected with the antiserum fast enough. The snake is immune to its own venom, just as predators possess immunity to the toxins in the prey they catch. Everything is involved in a constant war of all against all. There are some ravenous bacteria out there that will eat anything, ranging from human excrement to industrial waste.

Unsurprisingly, our body has evolved ingenious mechanisms to help us to work out which bacteria we want to invite into our body. Given potential partners' clothes to sniff, first-time daters are drawn not to the sweat of those with a similar immune system, but an opposite one, on the basis that two genomes fight infection better than one.

Early protection
Building our defences starts early. 'Morning sickness' is a pregnant mother's natural defence mechanism, her body instinctively rejecting certain foods as a possible threat to the child she is carrying. We are born virus-free, but we pick up our first microbes as we pass through the birth canal of her vagina, then from her skin, kick-starting our immune system.

If we are breast-fed, colostrum in her milk is a brew of hormones and antibodies aimed at immunising us against the dangers of the outside world once we leave the protection of her womb. Bifidus in breast milk lines the gut with protective bacteria in a way that formula milk cannot, giving us time in our early months to grow a 'safe' microbiome based on our own acquired defences. Whether breast fed or not, our first thousand days are crucial in growing our antibody defences and building our own microbiome by being exposed as soon as possible to a varied diet of solid food.

Birth by caesarean section has saved the lives of many mothers and children in modern times, but some bacteriologists worry that babies delivered into the world this way do not pick up the important 'safe' bacteria of the mother's birth canal on the way out. In addition, those born normally are closeted in over-heated homes, scrubbed with 'antibacterial' soap and over-protected from 'good' germs. Some doctors recommend a cocktail of friendly microbes alongside regular vaccinations to immunise babies against diseases such as leukemia, ward off allergies and protect against type 2 diabetes.

Ironically, good hygiene might be part of the problem, not the solution. Babies must be allowed to crawl around on the floor, because normal household bugs prompt the production of a wide range of antibodies, without exposing the child to life-threatening disease. Common diseases cause short-term distress to a child, but there are positive long-term advantages to priming their immune system early, as we have just seen with measles.

An infection-free childhood sounds a good idea, but in adulthood it can lead to dangerous allergies and auto-immune disorders, which have increased fourfold in the last forty years, though no-one knows why. Anaphylactic shock is a particularly severe reaction to what the body perceives as a threat, even though it might be only a few molecules being released by someone opening a bag a peanuts nearby.

There is increasing evidence that first born children have a weaker immune system than later siblings, and more allergies. They were involved in a tussle with the mother's antibodies in the womb, but once the battle has been won, later brothers and sisters are able to absorb the benefit stored in the lining of her womb.

Young children need all the protection they can get, because despite our better medical knowledge, there are many new threats: powerful toxins in our environment, high stress levels, superbugs, living beyond the age our bodies evolved to cope with illness, and a medicine cabinet of pills that freeze our immune system into permanent 'off' or over-activate it to permanent 'on', meaning it can end up attacking the very body it was designed to defend.

Even taking a painkiller can weaken the body's resistance to certain harmful bacteria.

There are some simple ways we can help ourselves to fight off infection. Exercise stimulates our white blood cells to stay alert, while strong bones and muscles are important factories for producing immune cells. A balanced diet feeds our gut flora, especially pulses, fibre and fermented foods such as yogurt, sauerkraut and kimchi. Perhaps the greatest incentive to staying healthy is that one of the most dangerous places to pick up a nasty infection is in hospital.

When the body attacks itself
Our body's instinct is to protect what is 'self' and attack what is 'other', but it is always caught between the safe haven of self-protection and the dangerous rocks of self-destruction. When under threat, our hypothalamus triggers a fever, raising our body temperature to burn off disease agents. This might make us feel even more ill, but it gives our immune system valuable time to mount its defences.

The 1966 film 'The Fantastic Voyage' shows what a titanic struggle this can be. A mini-submarine, sent into someone's bloodstream to repair the brain, is met with fierce response from an army of macrophages, literally 'big eaters' that are sent out at the first sign of trouble to devour the opposition. Meanwhile the body goes into lockdown, the lymph nodes swelling to prevent infection getting through the road blocks.

Normally the home team wins, sending out specially programmed 'killer cells' made in our spleen and bone marrow to seek and destroy specific pathogens. Sometimes these can get over-eager. In their enthusiasm they produce cytokines, or toxins created in the fallout of battle. Usually the body clears these away, but if they linger in the system, they can cause more damage than the original infection. For victims of Covid 19, it wasn't the presence of the virus in their lungs that led to breathing problems, but the inflammation caused by the waste products created by the body's attempts to eject the alien from its system.

Normally our auto-immune system scales down its 'red alert' response after an infection, but there can be severe organic and functional complications when it over-reacts, or stays in emergency mode, producing inflammation throughout the body, in places where there is no threat. This includes the brain, where cognitive function and memory can be impaired. An inflamed mind can result in depression through a shrunk hippocampus,

146

inhibiting the production of new cells, and a swollen amygdala, activating old fears.

Worst of all, the body can end up attacking itself, 'good' cells being misidentified as 'bad' ones. The condition of lupus is caused by the body's over-production of antibodies, which fail to distinguish between self and not-self. As a result, what is of benefit is treated as harmful, and what is harmful is allowed to flourish. Less life-threatening complications might be allergy, anaphylactic shock, asthma, coeliac disease (gluten intolerance), chronic fatigue (ME), dementia, diabetes, colitis, depression, obesity, multiple sclerosis, psoriasis, rheumatoid arthritis and sepsis, not to mention cancer and lymphoma.

Many of these conditions have a genetic component, perhaps once serving as a useful adaptation in our evolutionary past, but there is a suspicion that increased urban living, or over-protection from common muck, has weakened our overall fitness and resistance.

Cancer, of which there are two hundred types, gets into the body by tricking the Praetorian Guard of highly specialised search-and-destroy warriors called T cells. These identify mutated cancer cells as aliens, but the clever invaders get through by disguising themselves as friendly. The Holy Grail for immunotherapists is to revive the power of the auto-immune system by re-teaching T cells to recognise and attack marauders, especially breast cancer cells which are masters of disguise.

Psychoneuroimmunology is based on retraining the body to recognise cunning interlopers. Apart from being a long word, it is also smarter than chemotherapy drugs that blitz the whole system in search of their target. CRISPR, or gene editing, might prove useful in giving our immune system a head start by eliminating faulty genes at the embryo stage. Such therapy is justified by its ability to relieve suffering at an individual level, but altering the germ line of those not yet born raises ethical issues that wider society has yet to consider.

A healthy gut flora
The guardianship of our body is greatly aided by our microbiome, or flora of stomach bacteria unique to each of us, as essential to our health and happiness as any possessions we own. Our microbiome can be compromised by stress, poor diet or lack of sleep, so dieticians, doctors and psychiatrists are paying increasing attention to the connections between 'bad' gut bacteria and inflammation throughout the body, possibly even disturbance of the mind. A link has been found between harmful strains of E.coli

bacteria and colon cancer, and those with mental health problems often suffer from intestinal irregularities.

After illness, one of the most important things is to get our gut bacteria 'right' again, as recovery will be delayed until we do. A balanced microbiome keeps our immune system strong, acting as a second brain. The neurons in our gut are renewed every two weeks, a reflection of how hard they work balancing what we throw down our gullet with our overall health needs. Their importance is felt when we feel 'out of sorts', or allow harmful bacteria to colonise our digestive system. As if to prove the point, the bacteria that cause gum disease have been linked to the onset of Alzheimers.

Many animals eat their own placenta, vomit and faeces, not because they are starving, but because this keeps their gut protection system topped up. This might explain why children raised on farms, close to animals and regularly exposed to dirt and poo, often grow up with fewer food allergies and bacterial infections.

We are not totally at the mercy of pus, poison and disease. There is much we can do to keep our gut bacteria and immune defences in good form. Some days we might feel sixty, even though we're only forty five. Just as we can have a biological age that gives the lie to our chronological age, so we can keep our immune system young by judicious life choices. It weakens as we age, but if we look after ourselves, there's no reason we can't enjoy the immune response of a person twenty years younger as we enter our senior years.

Overuse of antibiotics
Although antibiotics work effectively in the short term, over-prescription can destroy our delicate microbiome, leaving us less resilient in the long term, as well as giving the green light to infections such as thrush and herpes, which a healthy body normally holds at bay.

Also, monoculture of both animals and plants has allowed antibiotic-resistant superbugs to flourish. As a result even more antibiotics are pumped into the animals we slaughter for food, which are then washed into our water supply. A test of the water in the River Ganges showed it contains more antibiotics per cubic centimetre than the human blood stream.

As a result of over-use, many essential antibiotics such as penicillin, first discovered by Alexander Fleming in 1928, are losing their potency for medical purposes. Our hospitals are infected with deadly superbugs, our oceans are infested with man-made pathogens, and after years in retreat, sexually transmitted diseases such as chlamydia, syphilis and gonorrhoea are making a comeback. Covid 19 showed in 2020 that there is an endless stream of different types of coronaviruses against which we need to be constantly vigilant, and for which we have no vaccines.

Research with the chemical excretions of bacteria in ants' nests is holding out hope of finding new ways of fighting infection. Phage therapy involves 'training' viruses to attack harmful bacteria, and engineering antibodies to seek and destroy cancer cells, which are expert at hiding among healthy cells. There is always the danger however that, unless these assassins are carefully targeted, they also blitz our 'good' bacteria in the process. For admirers of acronyms, there is also SNAPPs, or structurally nanoengineered antimicrobial polypeptide polymers.

Our endocrine, immune and nervous systems 'talk' to each other so intimately that even a placebo antibiotic can depress symptoms or stimulate resistance. This may matter in future, because pumping our body and environment with foreign substances has compromised our auto-immune system to the point that it might fail to spot intruders, remain on fixed alert or even attack itself, leaving us vulnerable to chronic inflammation, because the natural communication between the brain and pituitary gland has been scrambled.

The breathing brain
Some species of turtle can go three months without a gasp of air, and frogs can slow their metabolism to almost breathlessness during winter. In our case, our brain needs a constant supply of oxygen. It burns up nearly a fifth of the body's intake, even though it is only a fiftieth of the body's mass.

In a curious confluence of life and death, some seek extreme sexual arousal by depriving their brain of oxygen as long as they can by self-strangulation, but this is a dangerous practice. The price of a more intense orgasm might be not just loss of consciousness, but death by suffocation.

Plant life did not begin to make the Earth's atmosphere breathable until about five hundred million years ago, to the point that death by drowning soon results if our mammalian lungs are denied oxygen-rich blood reaching the brain. This can take several minutes, but after this time the damage to brain cells is irreversible, the same as having a blood clot in the brain caused by a stroke or embolism.

A case is on record of a baby boy surviving up to seven minutes in freezing water before rescue. His brain went into 'shut-

149

down' mode, making recovery possible without long-term damage. This is possibly a vestige of a mammalian diving reflex, taking blood away from the body surface to ensure the brain is not starved, but such instances are very rare.

Altitude sickness, especially when combined with exposure to prolonged cold, can prove fatal to the tired mountaineer. It slows down the heart, denying the brain vital oxygen, as a result of which the body enters a progressive shut-down sequence of fumbling, mumbling, stumbling and tumbling. Those who have lived to tell the tale report how the brain hallucinates, its abilities to process thought going into reverse, making it very difficult to decide whether to press on or turn back, a crucial call that cannot be made when optimal brain function is seriously impaired.

We don't have to climb a mountain to demonstrate to ourselves the importance of oxygen to our automatic body functions. We tend to take our breathing for granted, taking over twenty thousand breaths a day, but when we focus on our breathing, we become aware that deep, rhythmic breathing has the power to calm us, bringing our body and mind more closely in tune with each other, a truth long known to practitioners of pranayama, or breathing-focussed yoga.

When we breathe through our mouth, our breaths tend to be shallower, but when we breathe through our nose, we activate our nasal breathing cycle, which gets oxygen more efficiently to our brain. Holotropic breathing therapists claim they can improve our physical, mental and emotional state, but changes in our breathing habits cannot cure serious illness.

Taking a deep breath before we embark on any decision or action improves our chances of success by priming our brain for a better performance, focusing our concentration and increasing our attention. When we breathe in deeply, our heart rate rises slightly, and slows as we exhale. When we take a series of short sharp breaths, we can hyper-ventilate, giving ourselves a mini-high, though this might result in distress for asthma sufferers.

These instances remind us that, in normal health, our brain and body form an integrated system in a complex geography of feedback. We now proceed to investigate the geography of the brain itself a little more closely, especially how it maps thought and feeling in a single landscape.

Exploring the mind's geography

How do we map the brain?
Neuroanatomy – neuromapping – representation – reality – global awareness

- Neuroscience offers us a map of the brain, though to achieve this, the mind must be its own map-maker.
- Modelling the brain is complex, because some functions are localised, while others are distributed throughout.
- The layout of the brain is more the result of evolutionary serendipity than custom-design.
- Each dimension of the brain gives different insights into how the mind works.
- Depth, or front-to-back alignment, poses the question of depth of field: how does the brain make a coded representation 'in here' of the reality that is 'out there'?
- The dimension of height tricks us into thinking that some brain operations are 'higher' or 'lower' than others.
- The dimension of width alerts us to the fact that we have two brains sitting side by side.
- This requires some complex neurology to unravel, as well as tempting us to think that there is a 'right' way of doing things.
- The brain also exists in the fourth dimension, created by time in time to ponder time's mysteries.
- It has been built from the bottom up, with a 'new' cortex on the top, but it is a mistake to think that we have left our reptilian past behind. It is still very much part of our mindscape, and we would die without its ministrations.

151

We Are More Than Our Brains

Getting our bearings

When we go on a journey, we like to have a map of the lie of the land, even if this map is largely in our head, or on our satnav. To have a map of the brain, we must ask our brain to be its own map-maker, and the explorer of its own geography.

A more appropriate metaphor than a map of our brain's internal layout might be a floor plan, seeing our brain as a house of many rooms where different things happen. Over a hundred years ago Korbinian Brodmann divided the cortex into numbered areas. His attempt to peer at the geography of the brain beneath the skull was a great advance on going by the external lumps and bumps of phrenology, and was remarkably prescient. Modern neuroscience confirms his brain map, especially the vital association areas that act as a 'grand central junction' where all the other areas are drawn together.

Brodmann's estimate of fifty-two separate brain areas was too low: MRI scans put the figure closer to three hundred. In structure and function, our brain is a highly differentiated organ. It is not a nondescript lump of meat, but the seat of a manifold mind, not only multitasking biologically, but also existentially generating our sense of *what it feels like* to be a private person in a social world.

Regardless of how many areas into which we divide the three dimensions of the architecture and geography of the brain, we have to remember that it also exists in a fourth dimension of time, with an archaeology and a history. It is the product of the arrow of evolutionary time and the flight of a human life, and therefore, like all our knowledge and understanding, it is subject to change.

The idea of the brain working on a vertical plane or spiritual ladder goes back to ancient times. Plato talked of our reason perched on top of our will, which in turn squats on our emotions. This was more an elementary 'folk psychology' aimed at illuminating human nature than an insight into how the brain is organised, but its influence lingers. We still talk of 'elevated' thoughts and 'higher' ambitions. but 'base' feelings and 'deep' traumas.

At the start of the modern age, Sigmund Freud echoed this ladder of ascent with his notion of an id or unconscious hidden in the brain's limbic or lower regions, a conscious ego living half way up in the cortex, and a controlling superego in the neocortex, popped on top as if an evolutionary afterthought.

Freud was not far off the three-layered bruschetta identified by current neuroscience. The crusty base is our limbic region, the seat of memory, emotional processing and the amygdala. The tomato passata is a region called the insula, coordinating language in the temporal lobe and sensation in the parietal lobe. The topping is the cortex, controlling body functions in the hypothalamus, and thought in the frontal lobe.

However we eat our snack, the brain has been diced and sliced in all three dimensions. In the vertical plane, some of us are typecast as top-down dreamers who have grandiose ideas, while others are bottom-up pragmatists who get to grips with the detail. The first group have the ideas but seldom enact them; the second get the job done, but are a bit dull. Horizontally, we describe ourselves as predominantly left or right brained thinkers. From front to back, we talk of having an idea in the forefront of our mind, or fishing for a forgotten memory in the back of it.

These divisions are only pictures or metaphors. They shape our perception, frame our psychology and determine how we experience the world, but they are not neuroscience. However we view them, width, height and depth don't account for how our brain works, because it operates as a unity, its activities spread across the whole brain in an interlocking network, which is yet another metaphor. We say the brain 'maps' the world in specific sensorimotor regions, develops 'pathways' and operates in virtual 'fields', because figurative language is all we have to describe what goes on between our ears.

The Swiss Army knife model

So here's another metaphor: the brain is a Swiss Army knife. It has many blades, each evolved for a specific purpose as need arose, often repurposing materials already in the tool box. This gives our brain versatility, but also allows in a lot of gremlins. Like the QWERTY keyboard, our brain works, but it was designed to solve a problem that has long since disapperared. Conflicts of interest abound, as we discover when our body tells us to go to bed, but our will insists that we stay up to see the end of the midnight movie.

We Are More Than Our Brains

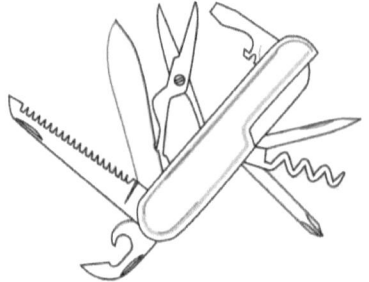

Swiss Army Knife model
According to this theory, the brain evolved one blade at a time. This means that, instead of being a smartly engineered super-tool, there are many design faults and conflicts of interest.

Each blade is a 'faculty' of the kind first suggested in 1768 by William Thompson, such as perception, memory, imagination, taste and judgment. But the central riddle remains unanswered: how does our brain unify so many separate functions and project them onto the whole body? Inside our brain there is no corridor of rooms with 'Head of Faculty' written on the doors. The brain does not work behind closed doors but is open-plan, integrating all its activities and networking all its operations.

This is because the brain is not a product but a process, a verb and not a noun, always in flux. Looking for its construction implies an architect with a plan, purpose and control, but time, opportunity and uncertainty have been the serendipitous designers of our Heath-Robinson neural landscape.

Evolution works within the contours of chance and necessity, each species 'bound' to navigate its path along tracks long laid down, but also 'free' to adapt to the evolutionary niche it finds itself in. Our brain has exploited this potential to the max, but not necessarily in a neat and ordered way. In the evolution of our cortical orienteering, new software has often been patched badly onto old hardware.

Mapping the brain

We must not take the idea of mapping the brain too literally. In the natural landscape, mountains tower over the valleys below, but in the brain, the later-evolved neocortex, though sitting on top of the brain, is not in any sense 'superior' to the ancient limbic brain that it

154

wraps itself around, nor does reason lord it over emotion, despite Plato's teaching. Because our brain works as an integrated system, there is no sense in which Freud's ego sits below the superego but above the id, and there are no neural stacks that ring-fence a region called 'I'.

The brain's geography is far more complex than any ordinary map, which gives only a 2-D view of the landscape. The brain is better understood as 3-D overlapping 'modules' that have evolved to deal with people, things and abstract ideas. These modules have been laid down over millions of years, but are so nested inside each other that it is all but impossible to tease them apart.

Mapping the physical brain tells us only so much. If we place a brain on a slab, intent on analysing its physical properties, we can see that it has the mass, texture, shape and appearance of a ripe cauliflower. It is ovoid, not cuboid, which is why we call clever people egg heads. We can calculate that, when in action, the brain consumes twenty per cent of the body's energy, and activates over half of its genes.

We have not however located the mind through our study of the brain as a physical object in three dimensions. We have to remember that our brain evolved not to sit in glorious isolation on a laboratory bench, but to perform as a social organ inside a living creature, jostling with other organisms for resources, attention, status and meaning. A solitary brain is an evolutionary impossibility. Without the cut and thrust of social exchange, desert island castaways eventually lose their minds. The body is of a piece with the mind, its meaning always residing outside itself, in its relations with other minds, and in the cultural patterns it is born into.

We will consider the brain's axes of height, width and depth separately, and the light each throws on our minding. Firstly however, we must reflect on two challenging questions that lie hidden in their contours: who is in charge, and how do we make it real?

Taking charge of the mind

The 'Numskulls' cartoon at the start of the book, based on a Beano comic strip, shows little men or homunculi busy in the ear, eye and nose departments, with a fourth homunculus reading a printout in the brain department, implying that he is the boss, or the guy who 'controls' the others and puts all the inputs together. But if this is what really goes on in the brain, it leads to an eternal regress: who controls the boss?

In the 1940's, the neurosurgeon Wilder Penfield discovered during an operation on the exposed brain of an epileptic patient that running a small current through a certain area on the cortex evoked a sensation in a corresponding part of the body. In other words, the body is mapped onto the brain. If that were not so, we wouldn't be able to touch our toes, because we wouldn't know where to locate them.

This suggested that the little men have fixed roles and abodes, as in the cartoon, establishing the doctrine of localism. But this is not the whole story. Penfield also discovered that touching particular areas triggered memories unique to the individual, suggesting that the brain also distributes information around the system. In other words, the little men are capable of extending their presence beyond their own departments, or of being in several places at once.

It remains a mystery how the brain achieves this paradoxical feat of fixity and fluidity. Neuroscientists have not discovered a commander of the fleet sitting on the poop deck. The claustrum, insula and hippocampus have been touted as possible command-and-control centres, but connectionism and network theory suggest that the brain is not an autocracy but a collegiate organisation, with distributed management and devolved responsibility.

As St Paul reminds us, we are many members, but one body. There would be no point in having a whole body given over to one sense, such as seeing. In any case, our eye would be much diminished without the other senses to feed information back to, or a whole body to guide.

Consider for instance the 'choices' that go on when we sit at the wheel of a car: eyes alert to the road ahead, pressure of hands on the wheel, foot on the accelerator, selecting a radio station, sensing how fast we are going, thinking about that problem at work. Fortunately we don't have to pay attention to all these inputs at once, because our brain prioritises: now it's the red light ahead, next it's the siren behind us. This explains why we can arrive at our destination with

156

no conscious memory of the journey. Our brain has 'driven' on autopilot, coordinating all areas subconsciously.

Cognitive neuroscientists have addressed the question of 'who is in charge' through the technology of neuroimaging, or shining a light inside the brain. This gives us a 'head start' of pictures of the brain in action that previous generations could not access. With electrodes on our head, we can actually 'see' our thoughts as they light up on screen.

The objective 'reality' of these pictures takes us no closer to the subjective 'reality' we experience as their owner. When we look publicly at another person's brain scan, we can see that neural activity is going on, and where, but we can't be party to what these thoughts are about, or what they feel like privately.

Making it real

How can these competing objective and subjective realities exist side by side? The answer partly lies in the complexity our brain faces in pinning each reality down, and then reconciling them. In order to map reality, it must first make a 'representation' of what it sees. This copy-input is 're-presented' in the sense that it cannot be the thing itself, only a neurally coded version of it. This means that when we 'think' of an object in its absence, or consult the map, what we see is a picture of reality, not the territory itself. The map is not the same as the territory.

On top of this complication, when 'we' read or interpret the map, we base our interpretation on our past experiences and prior expectations. This means we can hardly claim to be innocent in our seeing. Our consciousness intervenes: this is me, in this body, with my subjective assumptions, trying to pull my brain's representations together into a coherent and objective form.

Objective realists address this confusion by saying that reality is that which can bite back. If my toe hurts after I've kicked a tree, the tree must be real. My inside reality is shaped by what I encounter on the outside. But reality can be subjective too: what about that tree I dreamt of last night? Even in the daytime, I can believe that the tree in my garden ceases to exist when I stop looking at it.

Seen this way, the outside is a creation of the inside. When I am not in my garden, the tree exists only as an idea in my head. Reality

157

is therefore a construction of the mind, *my* mind, or the way *I* see things.

This view is called solipsism, and it has obvious weaknesses. I cannot dismiss whole forests or the logging industry as 'my' invention. Solipsism does not satisfy a tree surgeon who needs to get up close and personal to lop off branches. Nor does it satisfy a behaviourist, who maintains that, since we can't get inside other people's thoughts, the mind is merely what can be observed on the outside. We have to assume that someone standing next to a tree with a chain saw *probably* intends to lop off some of its branches. Finally, the moment two solipsists sit down and start chatting about the best way to prune a tree, their solipsism vanishes, because they are sharing each other's worldview.

No wonder philosophers of mind struggle to explain how the mind creates reality, or generates a self to own it. Here is me on the inside looking out at you on the outside, but when I peer into my brain, I don't see a chief executive sitting in a head office. So where does my sense of 'self' reside? Also, which gives me my surer knowledge, my senses that 'show' me the perceptible world, my feelings which manipulate my responses, or my reason which elevates me above such distractions?

Plumbing the depth
The front-to-back alignment of our brain gives our mind its depth, the first dimension we shall consider. It provides some answers, but throws up even more questions. Our eyes look out onto the world, as object, but behind them our brain has to make a model of reality 'in here', as subject. This sense of 'depth of field', an inside trying to make sense of an outside, is integral to our sense of being and knowing in the world, pitching us into tricky debates about perception, idea-formation, self-making and the construction of reality.

How our brain achieves the feat of cognitive world-making is a complex science, because our neurons store information as digital code, not images. We might *see* a picture in our mind's eye, or recall a sound in our mind's ear, but in our brain these are representations stored in neural codes that look and sound nothing like 'reality'. We get a glimpse of the difference between representation and reality when we go into 'systems' in our

158

computer. The inscrutable binary coding inside the machine looks nothing like the 'user friendly' image we see on the screen.

It remains a great mystery how our brain turns chemically coded information into technicolour experience and consciousness. Some claim that the time is not far off when we can build an artificial mind capable of the kind of inner life that we enjoy. In order to do so, engineers will have to replicate the techniques and specialised neurons that the human brain has perfected for making sense of the world, operating below our conscious awareness and control.

The cleverest minds in Silicon Valley are working hard on this, but they face huge challenges. For a start, the brain doesn't operate with a single neural code but many, with dozens of ways of talking to itself. Imagine for instance what would happen if an auditory message were delivered to an eye neuron or network: what would the ear see or the eye hear? Bizarre as the question sounds, that is exactly what happens in the brain of synaesthetes and the congenitally blind.

Another challenge for the designers of artificial brains is perception, or interpretation. We never just see, we 'see as' or 'see that'. Miss Muffet does not see a spider in the same way as an arachnophile, or a hungry blackbird. In ways that we never fully consciously grasp, our perception is primed. We see what we expect to see, or want to see. As a result, we'll never know what our computer camera 'sees' when we show it a spider.

Scaling the height

If the dimension of depth in the brain is phenomenologically complex, its vertical axis poses no less of an ethical conundrum. Moving *metaphorically* upwards from the bottom of the brain, we journey through eight million years of primate evolution towards the dilemmas of the human condition. This leads some biologists to see our brain as a kluge, or clumsy piling of one adaptation on top of another, the animal urges of our lower/old brain constantly making unwanted intrusions into the reasoned restraint of our higher/new brain.

As well as giving us a model of a triune brain populated by reason, will and desire, Plato saw the soul as pulled by two strong horses in a tussle for supremacy, one driven by raw passion, the other by noble reason. Today these steeds have been renamed

We Are More Than Our Brains

Limbus and Neocortex, and most brain researchers do not agree with Plato that they are always out to best each other. Feeling can be intelligent, and intellect can be irrational. It is far too simplistic to say that passion renders us stupid and reason makes us smart. Our wisest actions integrate emotion and reason in equal measure.

Nevertheless, Plato bequeathed us a dualism of 'higher' spirit and reason over 'lower' sense and passion, or idealism over materialism, which is perhaps why early thinkers believed mind and morality, elevated at the top of the brain above base desire, to be divine implants. Critics of such views say that this is not a very rational or top-brain thing to believe at all. A glimpse at any news bulletin reveals how little we have evolved beyond the emotional dictates of our lower brain, and how poorly our rational brains are succeeding in building a better, fairer world.

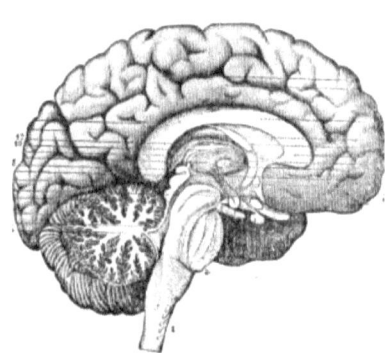

Peeling the onion
This cross-section shows the thought-controlling cerebrum, the most recently evolved part of our brain, wrapping itself around the mid-brain and brain stem, the seat of our feelings. Bottom left sits the cerebellum, or little brain, regulating movement and posture.

The brain looks not so much like a triple-decker sandwich but an onion. A series of later-evolved cortical or thought-oriented layers wraps around a central limbic or survival-focussed core, each successively added in response to need and opportunity. If we start at the centre and work outwards, we are in effect travelling through sixty million years of mammalian evolution, eight million years of primate evolution, and finally the doubling of hominid brain size about two million years ago, resulting in our dome-shaped neocortex. Anthropologists still debate the factors that led to the

160

We Are More Than Our Brains

Homo 'big brain', and the explosion of human culture two hundred thousand years ago.

This descent-cum-ascent explains why our species has been described as possessing the instincts of a lizard (limbic, reptilian or lower brain), the morals of a monkey (cortex, mammalian or mid-brain) and the cunning of a fox (neocortex, new or top brain). Some like to imagine there is an even higher level, the wisdom of an owl, sitting on top like an angelic halo, though few of us feel we can lay claim to such levels of illumination.

At the lower reptilian level, we might not find this characterisation of human minding very flattering, but when we look at the snake-like machinations of Shakespeare's anti-heroes such as Iago, Edmund and the Macbeths, we see ourselves in the mirror. Darwin was right to suggest that humans differ only in degree from animals, not in kind, because our brain has the same basic floor plan as a shark or a rat. And yet Shakespeare's villains are never less than human.

Encompassing the width
The third dimension, the horizontal breadth of the brain, presents us with a puzzle of a different kind. Inside our skull there is not one brain, but two, connected by a super-highway called the corpus callosum. How do two brains give us one mind?

We are made aware of our hemisphericality when a migraine knocks out half of our visual field. Bilaterality or asymmetry emerged as an evolutionary necessity about six hundred and thirty million years ago. As organisms grew in size, started to move and needed to feed, they had to develop a sense of a top and a bottom, a front and a back, and a left and right side.

In human evolution, laterality reached its apogee in the sophistication of the human hand. Primates rely on brachiation or the use of their arms to swing through the trees. When our Homo forebears came out of the forest onto the open savannah, their hands were no longer needed as hooks, and as they evolved upright posture, their hands were not required as additional feet for 'knuckle walking'. This allowed the opposable thumb to evolve, capable of touching each finger in a pincer movement, creating a manual finesse which no other primate possesses.

We Are More Than Our Brains

The upshot over many generations is that the human hand was freed for much more subtle operations, and this required upgraded neural control. Evolution decided that handedness, or giving one hand dominance over the other, was the best way forward, but this involved some radical rewiring. Like the braking system on a modern car, and to ensure that the body didn't operate as two unrelated halves, the left brain was given command of the right side of the body, and vice versa.

This means that the right hand is controlled by the left brain. Right-handed stone tools have been found dating back to one and a half million years ago, and some researchers link the rise of left brain dominance to the evolution of speech about a hundred thousand years ago in the left hemisphere. Chimpanzees show no preference for the right or the left hand, but nor are they capable of speech.

About ten per cent of the population might have been left-handers, as remains the case today, but genes and custom ensured that, for humans at least, tool-making became standardised for use in a predominantly right-handed world. To this day, screws are right-threaded, not as a conspiracy against left-handers, but because pronation, with the palm pointed downwards, allows for greater muscle force for turning and tightening with the right hand.

Where there is increased complexity, division of labour makes sense. When we listen to music, our right brain responds to the melody, but leaves our left brain free to focus on a conversation. Some find it easy to read while listening to music, even to be stimulated by it, but it is hard to concentrate on the text when somebody talks to us, because our left brain can process only one incoming channel at a time. A mother with a chattering toddler and a phone call to answer manages both, but she has to divide her attention to achieve it.

Despite our 'handedness', we don't experience the world as a divided reality. There is a cleft down the centre of the brain, but it is bridged by the powerful corpus callosum. This helps us integrate inputs and operations into a cognitive whole, synchronising 'phase differences' in messages that arrive at our ears microseconds apart. When we look through binoculars, we see only one image, not two as shown in old movies.

We Are More Than Our Brains

When we wear an eye patch or lose sight in one eye, we still see the whole field of vision, not half of it, because each side feeds its input to the other. The only time the brain sees only half the picture is after a stroke or brain tumour, destroying neurons in the visual cortex. Even after a commissurotomy, when the corpus callosum is severed for medical reasons, perhaps to prevent epileptic seizures, two halves of a brain still give a whole field of vision.

Solving the complex neurology of two brains making one mind is hard enough, but the cultural fallout is problematic in quite different ways. We can become snared in irrational 'leftism', believing that the left hand (controlled by the right brain) is odd, awkward and wrong, while the right hand (controlled by the left brain) is correct, good and....well, always right.

These biases are psychological and cultural 'constructions of reality', in no way attributable to how our brain is 'constructed'. It is too inter-connected to be reducible to its architecture, and it cannot afford favouritism. Also, laterality is reversible in nature. Fiddler crabs have a fifty/fifty chance of sporting their giant claw on the left of the right, and some of us are born with our heart on the right of our body, not the left.

Geological layers

The 'Schematic layers of the brain' tabulation below shows the layers of our brain in a vertical configuration. Our ascent through oblivion, desire and will to eventual reason mirrors not so much biological reality as an ancient myth of an ascent of the spirit. The soul, if it is lucky, progresses from demons in caves and wraiths in the underworld to a glimpse of gods in high places underneath an overarching firmament. Lofty mountains and tall towers have long been associated in religion with glimpses of the eternal, and pure reason has been revered as the gateway to the infinite.

Dr Jekyll and Mr Hyde might make for a good story, but the geography of the cranium is not so simplistically drawn. Unless we suffer from a particular brain lesion or organic malfunction, our personality is not split between goodness and sanity in the daytime, and wickedness and mania at night. We can't 'explain' psychopathy as the banishment of the rational Dr Jekyll to unleash the evil Mr Hyde, or meditation as the freeing of our neocortex from the distraction of cortical sensation and limbic craving.

163

Schematic Layers of the Brain

These 'layers' only hint at how everything in the brain is intricately interconnected. What matters is how we allow our 'higher' critical thinking to challenge our 'lower' prejudices, and whether we permit new 'bottom up' inputs to shake up our 'top down' thinking habits.

Neocortex forebrain plan the future extended consciousness	*200 thousand years ago*	**modern human** *rationality* - reasoning outside space-time *intellect* - 'cool' reflection *superego* - exercising control *cognition* - thinking abstractly
Cortex midbrain deal with the present self consciousness	*120 million years ago*	**mammalian** *reflection* - second-guessing what comes next *emotion* - 'hot' response *ego* - being aware *intuition* - relying on winning strategies
Limbic brain hindbrain cope with the past core consciousness	*300 million years ago*	**reptilian** *reflexivity* - reacting to the present *sensuality* - processing raw feels *id* - answering to basic drives *physicality* - living by instinct
Brain stem primitive brain automated processes zero consciousness	*2 billion years ago*	**earliest life forms** *gateway* to the brain *controller* of the nervous system *regulator* of body functions *prompter* of gag and startle reflexes

We Are More Than Our Brains

Emotion and reason are laced through every brain layer, so there's no such thing as a purely rational decision confined to our neocortex. Our reasons for choosing anything are always threaded through by our feelings towards it, as we realise in a job interview. In any case, all of our sensory inputs are first processed through the thalamus in our 'lower' brain, where they get emotionally tagged, so even our purest idea can't shake of its limbic origins and bodily beginnings.

If we think of our brain as an iced bun, the later evolutionary layers are like sugar sprinkled over the top. The sugar has however soaked right through the bun, down to the bottom, allowing us to be neocortical and limbic in the same moment. Also, we're not just sweetly reasonable on the top, or on demand. We can only be reasonable about something in particular, which means we can never bypass our feelings.

Neuroscientists see the brain not as a layered cake but as a constant mixing of ingredients. The only exception is when an injury damages a specific network, resulting in the loss of particular abilities. Even then, other functions can remain unscathed. The neurologist Oliver Sacks tells the story of a patient he treated who had a malfunction in the face-recognition module in his brain: he could not distinguish between his wife and a hat. But despite this very specific neural disability, the same patient was still able to play a minuet on the piano.

Memory, pleasure, feeling and reason are not holed up in dark corners of the brain, but spread throughout. Reason is not an 'essence' we are born with, but something we learn, as the by-product of many thinking activities that improve with practice. This flexibility is good news, because it is what allows our brain to reroute and reorganise after a stroke or injury of some kind. In a healthy brain, our memories, emotions and thoughts, even our 'gut feelings', work globally, permeating the whole of our conscious awareness.

Our 'maps' of the brain cannot therefore be accurate depictions of the landscape, but explanatory models or schematic diagrams, like the map of the London Underground, which shows the *relations* of stations to each other, not their real locations on the ground, or distances from each other. If we remember that we are persons, not maps, and that each mind is uniquely formed, we will

165

keep models, schemas, diagrams and metaphors in perspective as only vague approximations of the truth.

So accepting the health warning not to take the 'structure' of the brain's mansion of many rooms too literally, we can now consider how our brain's bottom-up evolution affects our top-down awareness and daily exchanges with each other. Do I like you (limbic gut feeling), do you owe me any money (cortical memory), and how can I get you to pay me back (neocortical cunning)?

What lies at the core?

The brain stem – limbus - cerebellum – thalamus - amygdala – hippocampus – archetypes – the unconscious - instinct

- The brain stem is our proto-brain, the part that keeps us alive, and the last part of us to die.
- It supports the limbic region, or 'collar' on which our 'new brain' or neocortex sits.
- Like any collar, it can support us or choke us.
- The amygdala, a small organ deep inside our limbic brain, is a good example of this. It can exaggerate our fears, or spur us on to greater achievement.
- It is not the amygdala that makes us violent or psychopathic. Only people can make those choices.
- If we gave way to our limbic passions, society would be selfish, life would be nasty and sex would be brutish.
- Without our limbic passions, we would not fall in love, cheer for our team or devote ourselves to a cause.
- Freud and Jung differed over what monsters lurk in the basement of our brain.
- Neuroscientists look for flows of neural activity, not dark corners which don't show up on a brain scan.
- Either way, we cannot ignore the promptings of our limbus. They are not instincts, but whisperings which stalk our dreams, dominate our politics, and make us think there is a point to living.

The brain stem
The brain stem, at the junction between our spinal cord and brain, is the core of our biological existence. It is our 'basement' brain, possessed by the most primitive animal with a nervous system, two

billion years old, as ancient as the first neural net that formed to guide a worm through the primordial sludge.

Its operation is not however at all primitive. It is the heart of our reticular formation, a net-like structure that works as a two-way coordinator between our nervous system and its brain. It integrates balance, movement, muscle performance and our sense of body-world boundary. Without it we could not judge whether we are being tickled by a lover, prodded by an aggressor or chafed by our own clothing.

It is the controller of all traffic, the gatekeeper of pain and pleasure. Through our endocrine system it regulates our breathing, appetite, digestion, excretion, sleep, temperature, heart rate and blood pressure without so much as a nod from our conscious mind. It may well be the fount of our consciousness. It is our 'vital spot', a fact tragically demonstrated when an Australian cricketer died in 2014 after being struck low in the back of the head by a fast ball.

The brain stem is our primary life support system, maintaining optimum operating conditions, the first part of us to form in the womb and the last part to die, even if we have dementia. When we are lying in a coma, lost to the world, it is our brain stem that keeps us alive, albeit perhaps with the aid of machines.

We blink, swallow and sweat without thinking. We are conscious of our breathing only when we get out of puff or are called upon to blow out the candle on our birthday cake. As our primitive or proto-brain, the brain stem houses the basal ganglia, a series of checking mechanisms that automatically keep us alive by spotting danger before we are aware of it. These work as closed loops, well out of reach of our conscious control.

Limbic support collar
Wrapped around the brain stem is the *limbus*, Latin for neck collar. Like any collar, it can either support us or choke us. The limbic brain is a legacy of our reptilian ancestors, regulating our powerful drives to feed, flee, freeze, flock, fight or fornicate. This does not mean that it reveals 'the nature of the beast', though it is the seat of emotion and motivation. As we shall see, the executive brain power and civilising mind activity going on 'above' it could not happen without it.

We Are More Than Our Brains

In the 1960's, at the height of the Cold War and the threat of nuclear Armageddon, there was a fashion for calling our species the 'killer ape', driven by aggression rooted in our limbus. Evolution can't pile one layer on top of another without 'remembering' what lies beneath, so the theory went. If we scratch ourselves, we find a bad-tempered chimpanzee beneath.

The zoologist Konrad Lorenz worried that our rapid growth in the neocortex has given us technical mastery, but that our nervous system is still stranded back in the Stone Age. Today, some see the limbus as the driving force behind anger-based populism and fear-based right wing politics.

The limbus is indeed our most ancient connection to our evolutionary past. Our sense of smell feeds straight into it, short-circuiting our rational filter and accessing our memory and emotions directly, explaining why smells can so quickly arouse, disgust or seduce us.

'Limbic' is a term rarely used by neuroscientists now, because they have broken this region of the brain down into its core components, revealing its sophistication. They do not see it as an area that has some strange unconscious pull on us, but as integrated into the operations of the whole brain.

The cerebellum or 'little brain', tucked as it is at the base of our skull, is not the seat of consciousness, but it lays vital foundations for later-evolved brain functions. It has expanded in proportion to the growth of the cortex, not only helping us to coordinate our movements, manual skills and vocal abilities, but also paving the way for symbolic thinking, which is the defining feature of human intelligence.

Such cleverness started early in our primate past. Swinging through the trees requires rapid predictions of weight, distance and speed. This automatic future-guessing ability eventually gave rise to the ability to think the thought without actually making the leap, allowing our hypotheses to die in our stead.

The cerebellum is increasingly attracting neuroscientific interest. Not only is it much larger in humans than in other primates, housing the medulla, regulating our core functions, the pons, acting as a relay station, and the putamen, processing learning; it also contains more neurons than the rest of the brain, many of which are highly specialised and intricately branched Purkinje cells that connect to networks in the neocortex, enhancing forward planning, abstract thought and social interaction. It is therefore like a brain within the brain.

The seat of fear and cruelty
The cerebellum also accommodates the amygdala, named after a Greek word meaning almond-shaped. It might be small, but here we register emotion, generate aggression, store phobias and feel fear. We know it is very ancient part of our brain, because it 'clocks' anything frightening or disturbing seen for less than forty milliseconds without our conscious brain being aware of it. Courage amounts to controlling what comes up from our amygdala, feeling the fear but doing it anyway.

We want our amygdala to be in the Goldilocks zone, just the right size and working at the optimum level. It does not generate fear or fortitude willy-nilly, but works closely with the insula higher in the brain, not randomly but in specific contexts. We can train it like a puppy. It can grow into an attack dog of fear, trauma and disgust, or a guide dog for courage, enjoyment and trust.

If our amygdala is overactive, we might be ill-tempered or controlled by our fears, too aggressive to keep our friends or too timid to take a chance. Orphans rescued from appalling conditions in Romania had shrunken brains except for the amygdala, which was enlarged. A swollen amygdala is also linked to long term traumatic stress, and a tumour on the amygdala is sufficient to send someone on a killing spree.

If our amygdala is small or underactive, allied with an absence of certain neurons in our frontal cortex, we might be in the one percent of the population who show psychopathic tendencies. This does not mean we will go out and kill someone. Many of us carry the genes for psychopathy, but they are not necessarily triggered, and even if they are, our psychopathy may be prosocial, not aimed at harming anyone. It is believed that up to four percent of managers and executives show some symptoms of psychopathy, their low emotional reactivity enabling them to make tough decisions without nagging pity holding them back.

A small proportion of psychopaths are low in fear and unable to feel the pain of others, making them capable of acting without conscience, in a cruel, cold, charming and calculating way we rarely see in nature. The cause may be genetic, though there is evidence that children who are separated from their mother or principal carer for more than six months in their first five years suffer 'affectionless psychopathy', one of the symptoms of which is a reduced amygdala.

Psychopaths suffer from 'flat affect', feeling no remorse for those they harm. Their emotional deficit means they can plan their cruelty weeks ahead, or serve their revenge cold in order to maximise their own pleasure. They can be manipulative if not callous because, although they know what feelings are, they do not feel them themselves. Their capacity to hurt others is enhanced because they possess low empathy and zero compassion. They can recognise most facial expressions except the one that sends a signal 'please don't hurt me', which elicits sympathy in normal brains.

These neural 'blind spots' present the legal system with a challenge when a psychopath harms someone. If psychopathy is a brain disease, humane medical treatment is called for, but if motives and choices are in play, there must be retributive justice. Either way, certified and convicted psychopaths need to be kept away from the public.

Animals are occasionally violent, but always as a means to an end, never an end in itself, or out of malice. We don't punish a shark for attacking a surfer. Wild animals have a larger amygdala than domesticated ones. They need to react quickly and show anger more often, faced as they are with more threats to their life and challenges to their status. Once dominance is enforced, their squabbles quickly subside, so there is no need to harbour grudges or hatch evil plots.

This explains why lawyers distinguish between spontaneous and premeditated violence. Crimes of passion such as a lover stabbing a rival unexpectedly found sleeping with his or her partner flare up from the limbus, the amygdala shooting its signals directly into the emotional right brain, untempered by the regulating activity of the frontal cortex. The violence is over in a moment, but might take a lifetime to atone for.

A planned murder involves meticulous deliberation in the left hemisphere, computing how to turn motive into opportunity. Ways of escaping detection must be explored, and it might be best to wait until the victim is sleeping. Fortunately premeditated murder is rare, and most of us exploit the passions of our amygdala creatively, combining its urgent promptings and our useful planning skills to achieve prosocial outcomes.

Limbic liaisons
Also in the cerebellum are the thalamus and the hypothalamus. The thalamus is the brain's relay station, playing a key role in conscious awareness, sensory processing and sensitivity to pain. When we feel that every part of us is aching, we can thank our thalamus.

The hypothalamus is regarded by some as the brain of the brain, the minder of the pituitary or master gland of our body, which monitors critical levels of hormone release. This makes it the guardian of our internal well-being. Barely larger than a pea, with only fifty thousand cells, it is the overseer of our moods, appetites, metabolic processes, sex life, eating, sleeping, heart rate and breathing, interpreting light signals from our eyes to set and coordinate our different body clocks. Its vital role is best understood by considering what happens when its regulating powers malfunction: hallucinations, anorexia nervosa, obesity and narcolepsy, to name but a few.

It's the part of our brain that raises our body thermostat when we have a fever, in the hope that 'having a temperature' will kill off the microbes that are making us ill. Its strong connection with the cerebral cortex is demonstrated by yogis who use the power of mind to control their heart rate and body temperature through biofeedback. When we die, or are about to be declared brain dead, it is our hypothalamus that is the last part of our brain to shut down.

It's important that this naming of parts doesn't deceive us into thinking that they work in isolation. On the contrary, they always operate in complex networks, not only with other brain structures, but also between other minds. A brain in a vat is impossible, because the brain is not a machine performing discrete tasks, but a convoluted organism in constant feedback with its surroundings. Only an embodied mind can respond to particular life challenges.

In mammals, the limbic brain also contains the hippocampus, named after its resemblance to a small sea-horse. In and around it there are 'grid cells' which act as our internal satnav. When we go the shops, it is these place-locators and direction-finders that help us to reach our destination and find our way home.

We Are More Than Our Brains

The hippocampus's most important role is probably as the seat of long-term memory, without which the past could not be stored in sequence. It also decides which memories are worth keeping in the first place. Not all creatures have evolved this facility, because their social lives are less complex. Fish don't possess a hippocampus, which is perhaps why we accuse our goldfish of not remembering how many times it has swum round the bowl. Recent research suggests however that fish have more of an inner world than we think, and can remember faces.

Adding spice
Despite our limbic system's roots in our animal past, or because of them, we could not be human without it. It is what makes us fight for what we believe in, take risks, get angry, make passionate love, protect our own kind, attend political rallies. It can thrill us to the bone, gladden our heart, soothe our savage breast and stiffen our resolve. As our rawest contact with reality, our limbic system ensures that everything doesn't feel the same. It adds spice to our life through shocks, surprises, epiphanies and peak experiences, or the moments when we feel most alive.

The limbic area is where we 'have' our orgasms, not as a single mental event but as the result of a complex mix of chemicals routed around the whole body. Wet dreams, which occur during sleep with no genital stimulation, prove that orgasms are limbic affairs, made deep in the brain.

Worthy Victorians were shocked by Darwin's suggestion that we share a limbic or hind brain with other mammals, indistinguishable in appearance from a pig's brain, testimony to the animal inside us all. But it is a reality we cannot deny. We are still adjusting to its implications for integrating our thought and feeling.

Our limbus is the hinge between our body and brain, and its language is emotion. Our body 'talks' to us in many ways, often making decisions before our mind does. We feel passion in our breast, anger across our shoulders, fear in our chest and disgust in our gut. These raw feelings have 'salience', forcing themselves on our attention as a kind of alarm system that there is something going on that we should notice. They are a kind of thinking, but at a lower level of processing. Our gut contains five hundred million neurons, but our brain a hundred billion.

173

We Are More Than Our Brains

We need these extra thinking cells up top because emotion is rarely pure and nearly always mixed with conflicting signals. It arrives at our consciousness as primary reactions of fear, anger, joy and surprise, but it requires a lot of secondary cognitive processing to decide how we feel about the situation before us. Anger might for instance generate more complex emotions of guilt, shame, sadness or sympathy. Do we let rip or bottle it for now?

A deep breath and a moment's mindfulness allow us to calm the maelstrom down below, proving that our mind is not the body's slave in these moments. Both feeling and thinking depend on the heart and the mind talking to each other: we can't hear one without listening to the other.

We get better at controlling our emotions as we grow, but when we are two years old they manifest as tantrums. Even as adults we have nightmares. We get swept along in football crowds, lynch mobs and religious crusades. We fall prey to chronic trauma, and in cases of shock, abuse or dementia, we see the body reverting to basics, where all that matters is survival.

Living with the limbus

The limbus is therefore the strongest but also the most vulnerable part of us, the Achilles heel of the human condition. If it is damaged, we put ourselves at great risk because we lose our sense of fear. We end up incapable of telling the difference between pleasure and pain, unable to distinguish between making love to a person and humping a gatepost. If we don't learn to control it we become limbic lackeys, prey to anxiety, paranoia and conspiracy theory.

In our early years the basic needs of our limbus need to be satisfied, for food, shelter, warmth, touch and affection, so that we can make secure emotional attachment with our carers, and grow up able to pass this gift on to others. Our brain is not just a biological organ that has to progress from tantrums and dirty nappies to self control and potty training. It is also a cultural construct, shaped by the smiles, verbal sparring and rituals of the human family.

Courtesy of this sustained support system, which assures our limbus that all is generally well, we can progress around our seventh birthday to the age of reason as our frontal lobes assume some control. For this to happen to best effect, our limbic brain needs to be a settled structure, not a cracked foundation. In brief, we need to be smothered in unconditional love, which is not the same as being spoilt rotten. Our early experiences, exchanges and moral lessons shape our perception of the world as a safe and welcoming or frightening and dangerous place. They give us emotional stability and gift us values which last a lifetime.

If all we receive in our early years is rejection, psychological trauma, sexual abuse or physical violence, we face chronic limbic lockdown. Even as adults, a shocked limbus, being wordless and pre-verbal, can require decades of therapy to reset the emotional sensors and work the damage out of the system.

The limbus is our body-within-the-brain, which is why we talk about being frozen with fear, scared stiff and gut-wrenched. A biologist sees a shrunken thalamus, but a therapist sees a damaged unconscious where past hurt leaves us in a numbed present. Time heals some of the wounds, but there must also be a cognitive dimension: the conscious mind must slowly take charge of the unconscious residue.

There is no consensus on how best to achieve this. Primal Scream Therapy means what it says: don't verbalise the past, but crouch in a huddle on the floor and bawl out the hurt. Depth psychologists believe more strongly in the power of words, not screaming in a burst of intensive regression. They try to talk their patient back to the original trauma, as a kind of exorcism of the past, even though this risks reawakening it in the present. This approach chimes with parents who encourage their children to verbalise their feelings after a tantrum, aimed at giving finer control over the next emotional surge.

Despite the fact that the trauma might be wordless and sub-rational, such as a phobia of spiders, cognitive behaviour therapists rely on a more conceptual approach. There is little point revisiting past fears to see how they began: it is better to focus on changing perceptions in the present. 'Flooding', or being thrown into a den of spiders, is likely to freak out the patient and reinforce limbic gut reactions, but gradual exposure, using real or virtual spiders, allows for a steady reshaping of attitudes and reprogramming of neural responses.

We Are More Than Our Brains

The Freudian id

Sigmund Freud named our limbic dark side the 'id', Latin for 'it', the seat of our unconscious. Here reside largely unverbalised and repressed memories, fears and desires, motivated by the pleasure principle, constantly on the lookout for the next thrill or hit.

Freud is often wrongly interpreted as suggesting that we are in thrall to our unconscious, but in fact he urged the opposite: by getting to know it better, especially as revealed in our dreams, we give ourselves more control over it, but only if we subject it to the gaze of our conscious awareness.

This means dredging up the unpleasant stuff from our childhood that we thought we had shut in a box out of sight down in the lumber room of our unconscious. Freud believed the emotions work on a hydraulic principle, building pressure from below to bursting point. If we don't grieve properly for our losses as a child, we will never be able to express our feelings as an adult.

We might have blanked out those times we wished our parents were dead, and those nasty things that happened in the woodshed, but they are still there, and we have to confront them out in the open. We might think we have masked them with rational thoughts, wise judgments and free choices, but our hurt emotions, distorted beliefs and guilty fantasies still lurk beneath. We might think we have buried them deep, but they pop up unbidden in as much as forty percent of our mental activity.

We persuade ourselves that we are in control through our cortex, the home of our ego. Here we are regularly brought up against the reality principle, as a kind of self protection. We might fancy another beer, but then we remember what happened the last time we over-indulged. Then there is our duty to others: if we drink-drive, we put other lives at risk too.

Enter the superego, or conscience, residing in the front of our neocortex. We find the mental strength to override the powerful urges coming from below. Freud believed that this capacity for self-denial, however much it pains us, is what makes civilisation possible, but it can also leave us deeply discontented. Like the bad-tempered two year old, we have to learn to cut a deal with life.

The superego is our annoying parent who knows what is best for us, even though we hate to hear its nagging voice. By teaching us that non-stop instant gratification seldom pays off in the long run, our elders frame our adult morality. We have to be warned that pressing ahead with an activity that might eventually harm us is foolhardy.

On the other hand, we have to take risks, find out for ourselves, and occasionally come a cropper. The teenage brain *needs* an element of risk if it is to strike out on its own and attain full independence. Facing down a threat, having weighed the dangers, is the essence of bravery, the foundation of self esteem and the guarantee of autonomy. The superego can help our neocortex to overrule our limbic urge to run away. This is our chance to stand up for ourselves, not conform to the crowd. We decide that going to the rescue of a friend is the right thing to do, even though it might pose a threat to our own safety.

The Jungian ancestor

Carl Jung took a different approach to the limbus from Freud. He traced it back not just to individual childhood but to the dawn of humanity. He called it the two million year old ancestor inside us all, its origins going back *three hundred* million years. It is the seat of primeval passion and atavistic unconscious that Plato and Freud warned us about. This is where our 'shadow self' lurks. No matter how strong we appear on the surface, there is a weakness that always lurks, the dark side of ourselves we want to keep hidden, spontaneously erupting when we least expect it.

It is the 'lizard brain' that demagogues appeal to, with its gut reactions of fear and tribal instincts of hate. This makes it easy for them to urge us to switch off our rational cortex, vote with our viscera and close our minds against the 'out group' who pose a threat to our jobs, women and children. The limbus seeks power, and its surges of euphoria and rage can literally render us speechless if not stupid.

Our political discourse is particularly vulnerable to limbic manipulation. Liberalism is founded on the ability of reasoned debate to resolve our differences, but liberal politicians have been accused of underestimating the role of feeling in how we choose our leaders. Populist rabble-rousers know only too well the benefits of playing on people's fears, hatreds, grievances and prejudices. The limbus is a sucker for conspiracy theories, repeated half-truths, fake news and false promises.

We Are More Than Our Brains

Populism is a version of politics that sets the id against the superego, or the people against perceived threats or oppressors. The problem is that if we set suspicion, private need, gut-feeling and immediate interest over trust, critical debate, mutuality and longer term policy, we run the risk of seeing society and global cooperation as a winner-takes-all race to the finish line.

We need to be careful about locating the unconscious in the limbic region of the brain. It's simply not the case that our hatred dwells down below in the dark, while our love basks in the light of day: each of us is a complex mix of simultaneous love and hate, and love in any case has its dark side.

> **Carl Jung**
> 1875-1961
> *Jung proposed that our ancestral past has left not only genes in our body but also archetypes in our mind. Not all agree with him, accusing him of imposing Western concepts on cultures which do not think like us.*

Jung acknowledged this ambivalence in his theory of archetypes. Instead of seeing the 'dark matter' of the mind as an amorphous atavistic throwback, he believed it is a repository of timeless psychic wisdom. This is not lost to us, but held in what he called the 'collective unconscious', visible in the characters or archetypes that populate world mythology: heroes who prove themselves, tricksters who get the better of their enemies, wise witches who guard against evil, earth mothers who nourish, children whose innocence atones for the guilt of their parents, rebels who challenge orthodoxy, explorers who break new ground.

178

We Are More Than Our Brains

Some anthropologists believe there is no cross-cultural evidence of 'A Mother of All Minds' populated by universal archetypes. We constantly encounter them nevertheless in their various guises in our story books and on our cinema screens. Jungian therapists use these positive role models to help their patients to learn to live with the monsters and mockers lurking within, not to fear them, because that way lies guilt and self-hatred.

Sex in particular is demeaned if it is consigned to the limbus. It is said that an erect penis has no conscience, but the majority of men are not limbic sex-puppets or slaves to their animal impulses. They are able to override ephemeral desire, commit to long-term partners and make responsible fathers. From a moral perspective therefore, while it may be the limbus that proposes, it is the neocortex that disposes.

Neuroscientists have partly confirmed Freud's theory of hidden forces, and Jung's notion of atavistic inheritances, but by a different route. They agree that we are not the proud rational choosers we think we are, but they don't put this down to psychic secrets buried out of sight. Instead they point to algorithmic neural programs that do most of our thinking for us, below our conscious radar.

These subsystems are spread through the cortex, even the body, not confined to the limbus. We become aware of their automaticity when we daydream, sleep, meet a stranger or switch to autopilot. A frog is even more reliant on built-in programs: it can still swim even when its brain has been removed.

Even when we think we are thinking, such as deciding whether to go on holiday, or who with, we rely not just on experience but numerous default calculations about time, cost, risk, preference and likely benefit. These are high-level processes but carried out below conscious awareness. After we've decided, we're sure we've made the right choice, but we've no idea why. It's only after the holiday, when we've had that ecstatic or traumatic experience, that the memories, good and bad alike, end up in our Freudian unconscious.

Staying alive

In Dante's 'Inferno', Limbus was the first circle of hell where the unbaptised pined away. As if this isn't infernal enough, its original meaning of neck collar reminds us how it can weigh us down. But a collar has an ambivalent role. It can be a millstone round our necks,

179

trapping us in ancestral knee-jerk anxiety, or a lifebelt, keeping us afloat with an instinctive gut-feeling for what is right. 'Gut instinct' is generated in our limbic region, not in our stomach: that's only where we *feel* it.

The limbic brain's main concern is to keep us alive, whatever the cost, at least as long as the threat lasts. We have forgotten the fear our ancestors must have felt as they stalked game while predators were on the prowl, but their limbic anxieties switched off when they got back to camp. This is not the case with the anxieties that chronically haunt our modern consciousness: pandemics, tsunamis, immigration, terrorism, internet trolling, superintelligent machines.

Nursing mothers are familiar with the 'baby brain', a naked cry for comfort, food and warmth, and none of us is beyond a screaming regression to infantile rage when we feel threatened. It is the limbic brain that wakes us up at 4am convinced that the world is doomed. Not for nothing has it been variously referred to as a wise old owl, crocodile or watchdog, because it plays all these roles for us. Its origins lie in the mists of the age of the dinosaurs, which is why 'Jurassic Park' nightmares still haunt us every night.

During REM sleep, when the rational neocortex is deactivated, sub-rational limbic fears stalk our dreams, making a mockery of our day-time pretensions to be cool and civilised. In our waking hours, the limbic region makes us respond immediately, leaving regret till later. In the jungle, it pays to react first and reflect later, after we've avoided being eaten.

It doesn't follow however that the limbus totally controls our reactions, such as the instant and unwilled startle-reflex of a baby. In their early years, young children at first show no fear of snakes, so we cannot blame the limbus for our 'instinctive' snake phobia. Fear of snakes is something we learn from the reaction of adults around us, and in our culture.

Once we have learned it however, we react so fast that we cannot override it. Charles Darwin mentions an occasion when he deliberately sat near a snake behind plate glass. Even though he knew he was safe, he could not stop himself from flinching when the lunge came, whether the snake posed a threat or not. His limbic brain 'saw' the snake before his cortical brain could override his reaction.

We Are More Than Our Brains

This is why we find it impossible not to flinch when someone feints a blow at our head: naked instinct beats over-dressed rationality any day. People involved in accidents say that in the split seconds before a crash their minds felt crystal clear, which may be an evolutionary throwback to the need for decisive instant response and a concentration of the brain at its most basic level.

In a tussle with the higher brain, the lowly limbic brain usually wins: when we have a phobia or get stressed, we scream before we think. When we are swept along in sexual passion, it's the limbic brain doing the sweeping. When we get cold or scared, our goose bumps go into action regardless

Lovable and laughable

Without our limbic brain, life would be very dull and barely human, devoid of passion and excitement. Limbic urges remind us that we are alive, making us respond to the cries of a baby, but they also make us yell when someone steals our parking space. So while the limbic brain can make us lovable, it can also make us laughable, even downright dangerous. It can be a lifebelt one moment, a millstone the next.

Its upsides are our intensely human moments of joy and passion, but its downsides include our vulnerability to addiction, fads, infatuations, phobias and outbursts of temper. It also prompts our gut-reactions of fear, anxiety and suspicion of strangers, unconscious energies that might have served us well once, but now have to be repressed, only to break unwanted into our conscious moments.

In an important sense therefore, Plato was right: as we age, we do become more capable of reining in our limbic horse more tightly. Criminal behaviour decreases with age, and if we are lucky, we live long enough to get a little wiser.

Nevertheless, advertisers know that this is the part of our brain they must target if they are to appeal to our pride, vanity, insecurity and craving for attention. They know that the fuel of the limbus is dopamine, driving us out to forage and seek reward. There is a stimulus, a demand to respond, then gratification, the brain's simplest program operating at its lowest level.

So when we go out and purchase that new item of clothing, we're either having our buttons pressed, or we're buying on an impulse we can't explain. When we are asked as moody teenagers what we were thinking when we did that silly thing, the answer is that we weren't thinking at all: we were collared by our limbic emotions. This is the millstone round our neck.

But the lifebelt saves our life too: the limbic brain is our smoke detector, and seldom sounds a false alarm. Just as the craftsman learns 'rules of thumb' that come with experience alone, so our limbic brain gives us that ineffable ability to sense what is best for us, or when something doesn't 'smell' quite right. It warns us when we are in danger, and helps us to tell a genuine smile from a fake one.

Limbic activity does not require permission from our prefrontal cortex. It kicks in automatically as a reflex reaction, and yet it is what makes us human. We can transcend it, but we can never exclude it. It gives us personality, makes us quixotic, and colours our raw perceptions with our emotions. So without it, there would be no discussion of the beautiful game, classes in aesthetics or thank you letters for favours received.

Who lives at the summit?

The cortex – neocortex – emotion – reason – chakras – neuroconnectivity

- The cortex is the most recently evolved part of the mammalian brain.
- The neocortex, or new brain, gives humans their pre-eminence over other species.
- From the outside the neocortex looks like 'grey matter', but our real thinking is done by the super-connected 'white matter' inside.
- If reason resides anywhere, it is in the frontal lobes of the neocortex, from where the brain controls its global operations.
- Reason cannot however be separated from our emotions. There is such a thing as the intelligence of feeling.
- Since ancient times, the spine has been seen as an ascent from lower bodily functions to the crowning glory of reason and the soul.
- More recently, the power of the mind-brain has been attributed to its connectivity, with all parts working in harmony.

The cortex

On top of the cerebellum sits the cerebrum, which houses the much larger twin hemispheres of the brain. Here we find the cortex, meaning bark of a tree, wrapping the trunk inside. In the womb, it is visible as early as five weeks. It dates from about eight million years ago, shared by all animals that have to think for a living, such as where to find the best food or get out of a tight spot. When we are fully grown, our cortex takes up three quarters of our brain's volume.

We Are More Than Our Brains

Here resides our waking self, processing information more calmly to adapt to changing situations through opportunistic learning. If the limbus is a reflex from the past, the cortex is a reflection on what we have learned. If the limbus is the watchdog, sounding the alarm, the cortex is the officer of the watch, marshalling the defence. If the limbus is the rugged off-roader, the cortex is the smooth limousine we take out for a Sunday spin.

The cortex provides top-down co-ordination of the body parts below. Animals with their cortex removed cannot walk, nor decide what they are going to do next. It enables us to plan our next move, and prefer one choice to another. It also gives us essential 'windows' into how the world goes about its business: night can be separated from day, and causes can be linked to their effects.

The upward shift from the limbus to the cortex marks a step change on the evolutionary ladder. Scans show it operates as three main areas: the sensory cortex, processing incoming messages; the motor cortex, controlling our movements; and the association cortex, where we analyse our experiences, shuffle our memories, balance our options and arrive at our decisions. In other words, it is where we are conscious, or self-aware.

This sounds a simple enough statement, but the mere existence of consciousness is highly contested. As with St Augustine's attempt to define time, we know what consciousness is until we try to explain it, and without a working definition, we can't decide whether it exists in other animals, or we should program it into our computers.

Describing consciousness eludes us because we are never aware of ourselves thinking; we just think. Our brain's trick is to bind all of our mental operations into one package, but hide the join from us. Its effortlessness and automaticity makes us think there is a layer across the top of our brain where 'we' are in charge, but there isn't. There are only the fleeting impressions of this moment, which vanish as soon as our brain states change. Neuroscientists are still not sure where exactly the 'show' of our consciousness is choreographed.

No wonder consciousness has been described variously as our sense of what passes in our mind, sensations overlaid with memories, our filter of reality, our defence against sensory overload, our feeling of what happens, an unbroken stream, even an unusable concept for meaningful investigation.

Our theoretical definition of consciousness matters less than the practical uses we put it to. It enables our knowing to transcend time and space. Our thoughts can range backwards and forwards, into and out of things. We can store memories that are emotionally tagged. Unlike the gazelle on the savannah, quickly recovering from the predation of one of the herd, we think 'that's my friend you're devouring'. We are fully conscious, or capable of reflecting on our mortality: next time, it might be me in the jaws of the lion.

To be broad-minded, we have to accept that though we are still victim to basic animal instincts, we also possess a higher capacity to put our drives on hold, which is the essence of morality. We can reflect on our actions and evaluate them against alternatives, which is the spirit of free will. We can defer self interest and take on the view of the other, which is the soul of community. We can question our senses and deduce counter-intuitive laws, which is the foundation of science. We can digest 'raw feels' and convert them into concepts and theories about the world, which is the light of intellect.

The neocortex

Spread liberally over the top of the cortex is the neocortex or new brain, also called the cerebral cortex, taking up half of our brain space. This made a surprise appearance in human evolution about two hundred thousand years ago, elevating us to Homo status by giving us the planning and symbolic processing capacity to read the words on this page.

The cortex is common to all mammals, but the neocortex is an extra wrapping. It is tightly folded to increase capacity to receive information-rich and symbol-laden messages directly from our eyes and ears. Its power resides not in possessing more neurons, but in generating almost infinite connectivity. It can rightly be described as the part of the brain that talks to itself.

Other larger mammals such as primates, dolphins, whales and elephants possess a neocortex, as well as spindle or econoneurons. These super-cells are evolution's way of turbocharging the brain, increasing its efficiency and sign-reading capacity.

185

We Are More Than Our Brains

The human brain has gone one better by also evolving the capacity for symbol manipulation: we can substitute symbols for reality, creating worlds of mind. A chimpanzee might sign to us that it wants a banana, but it can't tell us what the phrase 'banana republic' means.

A chimp's brain will reach a symbol-making plateau a year after birth, but a human brain will see more brain-building activity in its first year of life than at any other time, absorbing sixty percent of energy intake, sucking in the world in great gulps. It will end up uniquely primed to extend itself into alternative realities, make connections, generate abstractions and transport us into infinity.

The neocortex gives us our legendary 'grey matter', but this accounts for only the top few millimetres coating its surface. Below that is our 'white matter', its paleness owed to the fat-insulating myelin coating the axons that reach up from down below. It is this white matter that gives our brain its superconductivity and high-speed processing capacity.

Children giggle a lot and cry easily because their prefrontal cortex controlling their rational thinking takes time to develop this thick canopy of connections. Up to the age of a year and a half, they cannot meaningfully be accused of naughty behaviour, as their neocortical neurons cannot yet process notions of right and wrong, good and bad.

By the age of two however, their neocortex lights up like a Christmas tree when they are confronted by a moral choice: I want to grab this thing, but I've been told that I mustn't. They are wrestling with some newly super-charged neurons that constitute their conscience.

Myelination is not complete until the age of five to seven years, boys taking a little longer than girls. Children respond to being socialised from an early age, but the delay in the formation of full cognitive function suggests that they should not be forced to read or undergo formal schooling too early, simply because their brains are not yet wired up for such tasks.

During adolescence their white matter will increase markedly in relation to their grey matter, giving their brain the enhanced connectivity and symbol-making capacity of adulthood. From the brain's perspective, learning and intelligence are not simply about how much grey matter we have. Much more important is the connecting up of our white matter, making an ever denser forest of myelinated axons.

In a sense, our neocortex never stops growing, because it is always learning, which means it is always reconfiguring itself. By our mid-twenties, we can justly hope that our large prefrontal lobes, containing three quarters of all our thinking cells, will have credited us with rationality and reflection. The higher faculties of reason, intelligence, imagination, language, conscience, emotional restraint and abstract thought will form the core of our personality.

The executive brain

The neocortex is our executive brain, giving us the capacity to formulate our plans and make our decisions. Most crucially, it is the seat of our morality, because it allows us to put thought ahead of action. By buying us time to avoid knee-jerk reaction and override our impulses, we can do the right thing even when it's the harder thing to do.

The neocortical brain is connected to the limbic system, and especially the amygdala, by an area called the insula. This has attracted increasing attention in recent years, because although the amygdala alerts us to danger, the insula moderates our response. Snakes are potentially dangerous, but it doesn't mean we can't handle or like them.

Our amygdala registers the pain of others, but our insula allows us to interpret it and experience compassion. It is the part of our brain that advises us who to trust, and how to convert raw 'feels' into higher feelings. Without these we could not, as the poet W B Yeats put it, feel capable of being gathered into the artifice of eternity.

The neocortex is visible in the first trimester in the womb. Once our cortex has processed raw sensation, our neocortex can convert sensory signals into concepts and emotions into feelings, starting to come into its own around the age of seven, the 'age of reason', by which time toilet training and temper tantrums become a distant memory.

Reason is therefore 'higher' than emotion in the sense that the cognition-heavy nerves from our eyes and ears feed straight into the brain, while the 'gross' messages of the body are routed up through

187

the spinal cord, giving us the feeling that, though impulses come from 'down below', tactics and strategy are decided 'up top'.

Our prefrontal or 'new' cortex gives us time to hold information in mind without rushing to act on it. We can hold fire on impulsive urges and consider that it may be better to delay gratification. It is where we locate our consciousness, and where we employ our subtle theory of mind, enabling us to enter the mind-worlds of others. Here is where we locate our sense of reality, our freedom, our deep thinking, our selfhood, in this space just behind our eyes.

In neurological terms, the neocortex is the 'hub' or magnet for all other brain activity, the residence of the chief executive who runs our body politic and all its ministries, synthesising the output of all the brain's mini-computers into an illusion of coherence. It is our seat of self, where we have a habitation and a name. To be liveable, life needs a self-directing person to live it.

The intelligence of feeling
It is a mistake to suppose that the neocortex deals only in reason. Emotion also needs to be intelligently processed as part of a balanced mental economy and a successful life. Patients with damage to their ventro-medial prefrontal cortex are able to understand their situation, but are incapable of reaching a decision because they lack emotion, and therefore have no value-system to prefer one option to another. They possess no intelligence of feeling. They are like Buridan's Ass, unable to choose between chomping on a mouthful of hay or munching on a carrot.

For centuries emotions received a bad press, dismissed as limbic distractions, but modern research is revealing what many of us know intuitively: there is a cognitive dimension to emotions, or an emotional intelligence that belongs to the whole brain, not just to its nether regions.

This means it is hard to fake emotion and get away with it. According to the James-Lange hypothesis, creasing our face into a smile sends 'I am feeling happy' signals back to our brain. This may work for a short while, but parading around with a false grin makes our face ache, and our friends soon see through our pretence. Emotion cannot be simulated, at least not for long, and most of us can tell the difference between a forced and a fake smile. A genuine smile arises as a whole-body response to a stimulus that engages both our feelings and our intellect.

This does not mean that emotion cannot be subjective or even unreliable. The film director Lev Kuleshov showed that we supply an emotional reading of a face depending on its surroundings. We read different emotions into an actor's impassive stare depending on whether it occurs in a horror movie or a love scene. Context is all.

Once again however, well-trained emotional intelligence, amassed through steadily accumulating experience, comes to our rescue. Everyday life, unlike the movies, is rich in context, and our emotional antennae are finely tuned. If not, we would struggle to know the difference between true love and infatuation.

To be the real thing, love must be not a temporary insanity or abandonment of common sense, but a reasonable choice and rational undertaking for a sustainable future, in a way that makes our sacrifice of self interest the considered thing to do. Were this not so, our understanding of sexual desire would be stuck at the stage of self-centred lust, unable to grow from romantic to companionate love, or from self preservation to other-oriented emotional commitment over time.

Teenagers need emotional intelligence as much as if not more than intellectual prowess, but it takes longer to grow. The adolescent brain undergoes a major restructuring, caught in a race between conquering the world and learning discretion, especially if it belongs to a male.

Teenagers are attracted to excessive risk-taking, discounting the future and making wayward decisions until full emotional intelligence, wise self-regulation and good judgement are attained in the mid-twenties. Our prisons end up being full of young offenders who didn't make it in time. One argument against locking up juvenile delinquents is that they are condemned to a life of crime before their neocortex has fully wired up. They are not yet mature or emotionally intelligent enough to reject it.

If there is one brain area that is our crowning glory, unique to human beings, it is the angular gyrus in our neocortical parietal lobe. It is not divine, but it allows us to make our advanced mathematical calculations, hold a conversation and work out what each other is thinking, as near to god-like qualities as we are likely to get.

189

We Are More Than Our Brains

Keeping the high ground

The rationalist tradition of the West is sceptical about there being a higher consciousness or transcendental level of awareness, hovering like a halo above the head in the religious paintings of a former age. Any mystical experiences we have are the products of unusual neurochemistry in parts of the brain originally designed for much more mundane and material purposes. Few lay claim these days to having visions, out-of-body experiences or glimpses into other realities.

And yet a feeling of transcendence is part of the human experience, not just a trick of our synapses. It is available to all of us when we feel part of something greater, perhaps on our own or in a crowd. It is a sense of flow, or being lost in the moment. Ecstasy, as a state of mind and not the drug, is a natural feeling of awe and serenity when we escape the confines of the moment, albeit briefly. It literally means standing outside ourselves, detached from our body, as if we are seeing things from above.

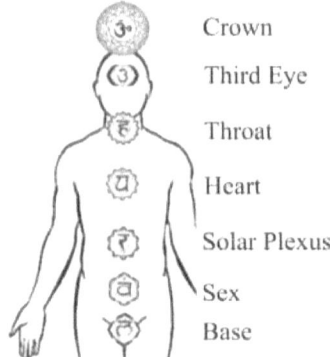

Crown

Third Eye

Throat

Heart

Solar Plexus

Sex

Base

The seven chakras
The ladder of the chakras takes us on a human ascent through raw instinct, habitual response, deliberate choice, creative intuition and reflective spirituality.

The Indian doctrine of the chakras offers the possibility of rising to a higher state of mind. These are energy centres or channels which ascend the spine like a ladder. Our feet are firmly placed on the lower rungs, but something seems to draw our gaze upwards, as if we contain all the stages of evolution within us. At the base is the root or sacral chakra (smell and sexuality), coloured red to symbolise instinctual survival. Similar to the limbic brain, the sacral chakra can be the source of security and contentment, but it can also generate fear of death and loss of control.

Moving up through the stomach (taste and will), heart (touch and love), throat (expression and communication), brow (sight and insight) and crown of the head (thought and transcendence), we progress from corporality towards spirituality, coloured purple to suggest its richness. This ascent mirrors our cognitive growth from the thought of a child to the *moksha* of a yogi, released from the ceaseless round of birth and death. This state of enlightenment is often represented as a third eye in Hindu art, symbolising a capacity to see truths and realities beyond what is presented to the mortal senses.

The chakras integrate the mind and the body, avoiding a dualistic split between thought and feeling. In the West we tend to assume that we think only with our mind, or higher energy centres, but the gamut of the chakras reminds us that 'being' engages our whole body and mind. High levels of stress, depression and alienation point to our need to move down the chakras as well as up, so that we do not lose our connection with raw experience and reality.

Things turn out badly when we privilege certain chakras over others. Fascists, dictators, populists and demagogues dress up their arguments in neocortical reason, but their real target is our limbic viscera or sacral chakra: tribal hatreds, sycophantic loyalties, bloody passions, primitive fears and naked self interest. It's fine to yell for our team (limbus), even to criticise the referee for his decisions (cortex), but essential to remember it's only a game (neocortex). Even politics has to be played by the rules.

If our brains are determined or 'written on' in any sense by evolution, it is at the lowly limbic level of temper tantrums, road rage and racial prejudice. Pavlov's conditioning experiments with salivating dogs exposed this knee-jerk part of our psychological make-up.

White people, shown a black face, often register an involuntary flutter of fear in their amygdala, until their neocortical reason comes to the rescue and moderates their prejudice, though never

191

completely. Given their poor treatment throughout history, black people in return might be forgiven for expressing a similar aversion to white faces. Once upon a time, all faces were hostile, especially different-looking ones, until proven to be friendly.

If we have any leeway to be self-willed, responsible individuals, overriding our desire for instant gratification, defying habit, controlling our urges, deferring pleasure, doing our duty, expelling prejudice, overcoming bigotry, converting stupidity to subtlety, experiencing transcendence, it is at the higher neocortical or crown chakra level. This can never be done in isolation. The head makes sense and assumes moral force only in relation to the body that supports it.

Straddling the contours

Most of the time we bumble along happily in the middle (cortical) until called upon to make a decision: hot desire (limbic) or cool reason (neocortical)? We experience the three levels as a unity or holarchy, each nested inside the other. We don't get up and say 'I feel particularly neocortical today', because our subsystems run through each other like the letters in a stick of seaside rock.

The limbic brain can function independently of the neocortex, as we see in screaming babies, but the neocortex can never entirely eliminate the limbus. Through a brain region called the anterior cingulate cortex, our limbic emotional activity and our rational cortex are constantly engaged in a game of neural snakes and ladders.

This enables us, as our neocortex develops, to learn how to contain our limbic tendencies, never able to put them fully behind us, as we discover in our dreams, frustrations and moods, when we make an evolutionary slide all the way back down to the bottom. Our limbic legacy is not so easily erased from our modern mindset. As a result, we experience our inner world not as a neatly layered three-tiered cake, but as a mixed loaf of perceptions, intentions, memories and desires, shot through in every mouthful with limbic, cortical and neocortical flavours.

In our teenage years, our neocortical frontal lobes are gradually rewired to inhibit impulsive limbic behaviour, giving us greater conscious control over our urges to show off, take risks and test the limits. We need time to find Aristotle's middle way between extremes.

Given that the brain is a work in progress at this age, mis-steps are inevitable, which is why most judiciaries around the world set the age of criminal responsibility as late as possible without undermining the concept of moral agency that slowly forms in the tighter networks of the neocortex. In the case of confirmed brain disease or malfunction, punishment is waived in favour of medical treatment.

As adults we need both limbic passion and neocortical reflection: when love bowls us over, we start to reason whether a relationship might be on the cards. The limbic brain wants immediate satisfaction, but the neocortex asks whether what we are doing is wise. To keep a relationship alive, love has to be both instinctual and cognitive: caring starts with thoughtfulness, another reminder that reason and emotion are complements, not opposites.

Biologically, the neocortex is a luxury and late-comer. We can get by without going to school, but we could not last a day without our limbic support system. We can survive damage to our neocortex, but malfunction of the limbic and cortical regions always proves fatal: we end up 'brain dead'.

Neuroconnectivity

The neocortex is divided into four lobes or regions. The frontal lobes have an executive function, integrating left and right hemispheres, sorting priorities bubbling up from further down, juggling with abstract reason. They are where we self-organise, pen our autobiography and keep ourselves in check. The temporal lobes govern sensory processing, linking hearing and memory to the limbic system. The parietal lobes give us a sense of being in our own body, and the occipital lobes process visual input.

The frontal and side lobes are deliberately not over-prescribed with specific tasks. This is not an oversight by our brain. On the contrary, it uses their open spaces and standby capacity for associative, integrative and creative thinking. Looking at them on a scan, we can see their connectedness, but only guess at the conversations they are having.

This means that we cannot sensibly nail down which parts of our brain do what, as if their functions are clear and fixed. If we do

193

so, we lose sight of our neuroconnectivity, the secret of our creativity, without which we would be unable to escape the straitjacket of sameness. It is our mind's extraordinary ability to generate a sense of wholeness and unity through vast feedback loops.

This is one of the great challenges for neuroscience: keeping sight of the whole mindscape. How does the brain integrate serial processing with networking, domain specificity with generality, biology with being, narrow functionality with broad-mindedness? Despite being divided into hemispheres, our brain manages to distribute information across the whole system through dendritic arborisation (tree-like branching) and reticulation (netting).

To change the metaphor, what characterises our minding is not a straight-cut canal system but a million rivulets coming together to form a mighty river, then breaking into a billion tributaries flowing out to sea. This cycle of inter-connectivity is nowhere more evident than in the integration of our brain's right and left hemispheres.

How are the hemispheres united?

The split brain – bilaterality – division of labour – asymmetry –
the corpus callosum – unity

- The brain comprises two hemispheres, necessitating some complex neurology: the left brain controls the right half of the body, and vice versa.
- We divide ourselves loosely into left and right brainers, but this can lead to unhelpful stereotypes.
- We characterise the left brain as breezy and the right as broody, but in the cut and thrust of life, this is a necessary balance and productive division of labour, not a recipe for conflict.
- In our evolution, two brains working in synchrony proved to be more versatile than one.
- Also, asymmetry can be a source of adaptive creativity.
- Our hemispheres are joined at the base by a super-connection called the corpus callosum, ensuring that we experience reality as a seamless unity.
- The corpus callosum is sometimes severed for medical reasons, leading to problems of 'one man, two guvnors'.
- In cases of disease or stroke however, depending on the nature and extent of the injury, one brain can deputise for the other.
- The links between hemisphericality and left/right bias in politics, if they exist at all, are cultural, not neurological.

Dual control
Biologically simple animals that don't have to go very far can get by with a single brain, but more complex locomotion necessitates bilaterality, or operating each side of the body separately. This calls for dual control, or a twin brain, working cross-wise as a back-up,

one side for each claw, antenna, fin, wing, paw or leg. Given this division of labour, walking without falling on our face is one of the cleverest things we do, and making a single sensible picture of the world from the input of two eyes is even smarter.

The psychology of having a split brain is much more problematic than the biology. It has become fashionable to talk of left-brain and right-brain thinking, as if more than one person lives at home. Our brain polarity has been touted as a master-theory to explain almost everything about us.

But we are not schizoid creatures forever doomed to a 'splitting' headache, or antagonistic thinking patterns. It is a neuromyth that 'warming up' exercises such as touching our left elbow with our right hand gets us ready for action or improves our concentration, because our biological survival depends on our brain doing this automatically.

It is also simplistic to accuse left-brain analytical science of stifling creativity, to dismiss the intuitive 'arty' right brain as over-emotional, to say that the left 'sees' while the right 'feels', or to blame a bad day on sleeping with our head on the wrong side of the pillow the night before. 'Getting out of the wrong side of the bed in the morning' is only a metaphor.

We simplify too far if we say that we manage our relationships with our people-oriented and metaphorical right, while we pay our bills with our calculating and workaday left. It does a disservice to our brain's capacity for integration to typecast the left as a dull pragmatist while the right is a creative neurotic, or to portray the left as a servant while the right is its master.

Like most human activities, language, number and creativity work to different patterns spread right across the brain. Watching a movie and reading a book trigger different types of 'picture-making' activity in both hemispheres, and surfing the internet can engage our 'rational' left hemisphere and 'emotional' right in the same moment. Any meaning we derive from the experience is an achievement of the whole brain.

We glibly stereotype each other as a left or right brainer, but evolution has deliberately made us wise enough to be healthily in two minds about most things. We are amphibious in our mental life cycle. As children we start out largely dominated by right brain feelings and intuitions, but we gradually narrow down to become creatures of left brain convention and habit. This does not mean that we are unable to retain our enthusiasm and creativity as adults.

Case histories

We gain some clues about the self-sufficiency of each hemisphere through medical case histories. It is possible to live and even pass university examinations with half a brain. Epileptics occasionally have a hemisphere removed to prevent life-threatening seizures, their remaining hemisphere capable of doubling up. At the age of forty five Louis Pasteur suffered a cerebral haemorrhage which destroyed a large part of his brain, leaving him paralysed on the left side of his body, but he recovered and continued working for another twenty years.

These cases prove that half a brain can operate as a whole brain, but the news is not all positive. Damage to the right brain can cause apathy and loss of belonging without the ordering of the left, while impairment of the left can cause delusional or manic episodes without the moderation of the right.

There are cases on record of patients with damage to their left frontal cortex caused by dementia or head injury who seem to develop sudden abilities to draw, paint, write poetry, compose music, handle numbers in novel ways, speak a new language, even develop a new personality. This is not so much evidence of a miracle as an example of disinhibition. The brake of the controlling left brain is slackened, allowing old memories to reappear, and freeing skills that had previously been blocked by other brain activity.

We might feel we instinctively vote for left or right politicians, but this owes itself not to brain structure, but to a search for a meaningful narrative. In any case, we are capable on any given day of switching sides, or holding opposing views simultaneously.

The mechanics of hemisphericality

Our hemisphericality raises three questions. Why has it evolved, how does it work, and how does it affect our experience of the world? Let's deal with the mechanics first.

Hippocrates was the first to realise the brain's bilaterality over two thousand years ago, noticing that if he touched one side of an exposed brain, the limbs on the other side of the body moved. Each of our eyes feeds its signals into both hemispheres, but visual processing of the left is concentrated in the right hemisphere, and vice versa. The prefrontal cortex closest to our eyes is always active (we react to bright light even in our sleep). This gives us a left/right division into two smaller brains joined by the corpus callosum. Smells are fed directly to both sides of the brain, possibly a reminder of how vital smell was to our early survival.

Whether we are right or left handed, handedness gave our species a practical advantage which contributed not only to the growth of our big brain, but also to language. While one hand is busy holding or grasping, the other is free to make signals. As a result, our brain finds itself increasingly multi-tasking and symbol-manipulating.

This practical edge leads to a set of cultural and psychological mind-teasers: how do two halves make a whole? If we are puppets pulled by diagonal strings, each half of our brain giving us a different 'take' on the world, how do we integrate them into a unified experience, and avoid privileging one side over the other? Also, if we have two demi-brains, do we have two sub-selves, or rival tenants in the building? As a thought experiment, what happens to 'me' if my brain is separated in two halves and placed in different bodies?

Apart from this transplant being beyond the bounds of medical expertise, the general opinion is that, in a normal brain, our language-dominated left hemisphere becomes the seat of our conscious awareness, the 'silent' right serving mainly as an unconscious back-up, so there would still be only one 'me'. We do not experience the world as two brains, even when we talk to ourselves. A healthy brain evolves to see and imagine the world from only one perspective at a time.

Continuity of identity
Our hemispheres give us coherence by sharing information, so we are not afflicted with a plurality of consciousnesses in one mind. Where such phenomena do occur, there is usually a neurological explanation, pinpointing a malfunction of normal brain behaviour. Ordinarily our hemispheres are not 'aware' of each other's operations. This drawing together of consciousness, or what

constitutes our sense of self, happens in our frontal lobes, in the place where we 'feel' it, right behind our eyes.

Rasmussen syndrome is a rare childhood disease in which one of the hemispheres causes seizures. In extreme instances it is removed, but the neuroplasticity of the young brain allows the remaining half to take upon itself most of the functions of the other half. In such cases, even a divided brain retains the essential unity of a *person*, a non-branching or indivisible concept allowing continuity of identity.

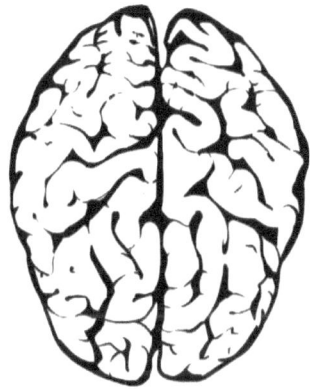

Twin hemispheres
The two hemispheres of the brain seen from above fit together like the halves of a walnut. They are joined by the superhighway of the corpus callosum, visible only from below.

Identical twins, though conceived from the same egg, go on to develop different personalities, because they have different life experiences. If our body could be cloned, an idea which remains in the realm of science fiction, there is no guarantee our mind would survive the transfer, and even if it did, our clone would not be privy to our private thoughts but soon develop its own. Even in cases of multiple personality disorder, where sufferers lay claim to more than one consciousness, perhaps to protect themselves from memories of childhood sexual abuse, essentially they remain one person, because their brain stem operates one beating heart belonging to a single organism.

The brain is so wired that its left hemisphere controls the right side of the body and vice versa, so if we clench our right hand, we activate our left hemisphere. We see this most clearly in how our eyes feed messages to each half. If our right eye is covered, and a funny image is shown to the left eye, stimulating the right hemisphere, we laugh without knowing why, because the 'explanation-giving' left hemisphere hasn't seen the image.

When asked why we are laughing, the left has to confabulate or invent a reason, to hold our sense of self together, because it is a know-it-all which cannot bear not knowing, and hates inconsistency. Desperate to turn ignorance into certainty, it sees connections between things even where there is no evidence of any.

A house divided
This is because our left hemisphere is our 'interpreter', story-teller or spin doctor which fabricates a narrative to resolve clashes or conflicts of interest between our inner world and outer reality, cobbling together fragments, resolving contradictions and filling in any gaps to form a single account of events.

The right brain makes the choices, then the left brain has to justify them. The right maximises the information, the left puts it into order. The right falls back onto assumption based on what has happened before, the left has to cover up any inconsistencies this may lead to. Then, once the left has settled on a narrative, the right plays devil's advocate, trying to expose the creaks in the plot.

This makes the 'self' sound like a story that we write about ourselves, a fiction arising from warring factions. We might balk at this idea, but we are caught in a strange loop that we can't escape from. Because our left brain is so clever at deceiving us, we can never prove or disprove it. The mind is not a see-through lens, but the very instrument of our seeing, filtering what is between us and the world. All we have to go on is our *intuition* that we are in control, which has evolved to guarantee us the truth.

Hunch alone provides insufficient proof that we possess a sovereign self. Consider how often our body language proves our duplicity. We hear ourselves spouting fine words with our left brain, as if butter wouldn't melt in our mouth, but the gestures controlled by our right brain reveal what we're really feeling. We fidget, avoid eye contact, give a false smile. Our guilt is written all over us.

Pragmatics comes to the rescue. Neurologically we don't operate with two brains, where the left hand doesn't know what the right hand is doing. Evolution has not been so careless. When we

200

take dancing lessons, we might at first feel that we have two left feet, but practice wires up new connections, and may even make us perfect.

So we're not normally aware of being a house divided against itself. Society demands that we tell the truth, but our brain's sole concern is to maintain neural unity and biological integrity. Any concept of the self has to embrace these two occasionally conflicting realities, weaving their separate threads into a single cloth.

The corpus callosum
To achieve such unity, our brain calls on the services of a bridge of two hundred million dense fibres connecting our hemispheres deep down at the limbic level, called the corpus callosum, so super-wired that we are never consciously aware of having two brains, eyes or ears.

We don't for instance say we prefer the images of one eye to the other, and by the same token, we never sense our hemispherical partnership as having two persons inside our head. We are not like the legendary Doctor Faustus who heard the voice of his Good Angel on one shoulder, followed almost instantly by his Evil Angel on the other.

But our hemispherical division of labour is real enough, as demonstrated by patients who suffer 'muscle-memory loss' if their right arm or leg is inactive for a sustained period, or affected by a stroke. They have to retrain the corresponding opposite hemisphere how to move that limb again, and to fire messages across the corpus callosum.

Experimental proof of this phenomenon came in the 1960's when Roger Sperry severed the corpus callosum in several animals to confirm that each half of the brain processes information separately prior to integration. He could not repeat this experiment with humans, but he could show objects differently to the left and right visual fields of male volunteers, screening one from the other.

When pictures of naked women were shown to their left visual field/right brain, hidden from their 'rational' left brain, they denied seeing such images, but a naughty smile appeared on their faces nevertheless. Sperry took this as proof that the so-called 'silent' right brain is not subordinate or speechless but has a mind of its own, with its own concepts and impulses to act. So the left brain doesn't call all the shots after all: the right brain rides shot gun.

Similarly, if shown a different image to each eye simultaneously, we have to choose one or the other, as we do in the one vase/two faces optical illusion. We cannot entertain a state of cognitive confusion. The only exception to this is the condition of 'visual neglect', in which one side of our visual apparatus is damaged, resulting in a lack of signal to that hemisphere. When asked to draw what they see, sufferers of visual neglect complete only half the picture.

The Stroop Test also alerts us to a clash of hemispherical functioning. If we are shown the word BLUE in orange letters, and asked to read the word as fast as possible, the colour-sensitive right sees the colour orange before the linguistic left has time to override it and say blue. Colours are primary stimuli to our brain, but words require secondary processing. Those who do the test are surprised at the level of concentration required to ignore the colours and read the words at speed without faltering.

This phenomenon reveals as much about executive control happening on both sides of the brain as it does about laterality. The automatic 'action' or system-one part of our prefrontal cortex in each hemisphere proceeds as normal until alerted to the need to change tack by the 'executive' system-two cingulate cortex, in charge of conscious intervention.

In other words, we stumble over reading the word because different brain systems are suddenly in conflict with each other, calling for an effort of will. Neuroscientists point out that the same brain areas are involved when we wrestle with difficult moral decisions.

When we watch a gifted keyboard player at work, we can but imagine the volume of neural traffic firing across the corpus callosum to coordinate two sets of busy fingers. Some jobs require high levels of cross-brain processing. Airline pilots, confronted with an array of display panels, need to be able to process two or more visual fields simultaneously. Most of us struggle to focus on one television screen at a time, let alone two. Practice can improve performance, and bilaterality can improve with age, perhaps because more pathways are laid down, and shortcuts are found.

We Are More Than Our Brains

In two minds

Occasionally, for medical reasons such as alleviating acute symptoms of epilepsy resulting from abnormal firing between hemispheres, the corpus callosum is severed. This can have interesting consequences, raising the question of which side of the brain is in charge, or is the seat of consciousness. Right-handed patients might suddenly find that they can draw much better with their left hand. It can also lead to patients being in two minds, or possessing two warring selves in one person, standing in front of a wardrobe with each hand making a different choice of outfit, or rejecting one of their hands as 'alien'.

In a rare condition called anarchic hand syndrome, left and right brain can be in conflict even when the corpus callosum is intact. In what must be a bewildering experience for the patient, one arm seems under control, but the other seems to have a mind of its own, acting randomly or inappropriately, as if controlled by a second mind.

The 1984 film 'All of Me' portrays this in comic form as one half of Steve Martin's male brain is taken over in error by a woman looking to enter another body, leaving him as half himself and half his unwanted guest. In reality however, even when our corpus callosum is cut, we don't become two separate persons, because to do that we would have to cut our brain stem in half too, and if we did that, we would die. So the whole thing remains on the level of an intriguing thought experiment or Hollywood movie, and nothing more.

At a functional level, we don't lead a double life, but we do perform a double act that results in a unified experience. We need both gloves to make a matching pair. When we listen to a stand up comic, our left brain hears the joke and decodes its literal meaning, but the right responds to tone, innuendo, incongruity and ambiguity, eventually 'getting' the punchline. We might not be sure why we are laughing, because the right brain can't verbalise its insights, but we'll find it funny anyway.

Because we can 'see twice' and infer 'double meaning', we can simultaneously see the funny and serious side of things. Our linear left brain does the syntax, but our allusive right brain sensitises us to the semantics of irony, paradox, sarcasm and metaphor, which colour and spice up (two right-brain metaphors) our appreciation of life.

Behind the appearances, the left is generally the logical side that controls analysis, language, rational thinking and cognitive processing, while the right is the emotional side more given to feeling, ritual, imagination and meaning in a wider symbolic field. We are conscious of what our left brain is up to, while our right brain's machinations are often unconscious and hidden from us.

The left gives us reasons for making a New Year's Eve resolution, while the right knows emotionally that we're not likely to keep it. The left tells everyone we are on a diet, but the right puts food in our mouth regardless.

Bright or broody
This bicamerality operates in dogs too. Their left hemisphere can recognise the meanings of up to a thousand words, but they rely on their right to detect the tone of what is being said to them. Their tail-wagging also reflects mood: to the right means happiness, but to the left means 'beware of the dog'.

Research in humans confirms this: our left hemisphere is more breezy than our broody right. If we feel happy in any way, it's thanks to our left brain. By the reverse token, depression may result from excessive speculation and moodiness in the right. There's even a suggestion that religious experience arises from the invasion of a right-brained 'sensed presence' into the matter-of-fact narrative of the left-brain. This occurs through a process called canalisation, as the left brain 'gives voice' to the wordless apparitions of the right in ways that the rational 'I' can handle and control.

These crossovers can be welcome, as the left can be a bit dour at times, reminding us of what we ought to be doing, whereas the right is more concerned with the experience of being in each moment.

Normally the left and right coexist in creative tension, but they can get out of kilter. Some therapists use eye-movement exercises as part of a treatment called neurolinguistic programming to help depressives stimulate and re-synchronise both sides of their brain, which seem to work, though no-one understands how. Imbalance might create other problems. If the reticent right is allowed to overpower and silence the talkative left, feelings can overpower thoughts.

We Are More Than Our Brains

Language helps us to articulate our feelings as a kind of exorcism, which is why adults try to get children to verbalise their thoughts, especially after a meltdown or misdemeanour. We stand a better chance of controlling our moods or working out our problems if we talk them through. Such relief is denied however if our negative right brain assumes control.

Keeping things in balance
Some commentators put the rise of populism of 2016 down to the fact that liberal democracy had become too left-brain analytical and rational, losing touch with right-brain desire for narrative and identity. Consequently, many who felt 'excluded' were resorting to the internet as an alternative outlet for their feelings, looking to connect with their emotional kin or 'tribe'.

This allowed them to bypass conventional 'wordy' political discourse altogether, but also put them at the mercy of 'sensationalist' fake-news trolls on digital media. Meanwhile, tabloid press articles stoked fears that 'they' were coming to take our jobs, steal our women, sponge off our generosity or jump our hospital queues. Whatever the causes, electoral results have since taken 'rational' pundits by surprise. Brain laterality might therefore be more closely connected to the idea of a political right and left than we think.

When we look closely, there is no neat template that places liberals in a left-brain haven of tolerance, high empathy, progressiveness and mutuality, while conservatives inhabit a right-brain realm of defensiveness, suspicion, tradition and self-interest. Where for instance would we place Josef Stalin in such a grid, a man who professed a belief in equality and social justice, but murdered twenty million of his own people?

Laterality might impact on our creativity, imagination and problem-solving powers. With its short and more tightly connected dendrites, our logical left brain gives us reliable and rapid analysis. Our right brain, with its more branching and loosely linked dendrites, is in less of a hurry.

We find therefore that if we change our activity, or sleep on a problem, like a scientist incubating a theory or a novelist hatching a plot, we might come up with more subtle solutions to problems. When insight comes, it feels like a 'flash' of inspiration, because a burst of alpha waves temporarily shuts down the distractions of our visual cortex, allowing more unusual juxtapositions of ideas to come to the surface.

By the same token, savants are reckoned to derive their extraordinary mental powers from a less active controlling left brain, allowing their right to blossom. There is also speculation that bipolar depressives in their manic or 'high' creative phases, especially artists and writers, enjoy the freer connectivity of their right brain, less constricted by their matter-of-fact left.

The key point to remember is not to take laterality too literally, or fall foul of the reductive fallacy, claiming that creativity or genius are 'nothing but' the effusions of an over-active right brain. Even if we insist that the left is a better linear processor than the right, or that the right is more tightly networked, the brain is far more connected than it is divided, and in some people the polarity is reversed. In the case of children whose left brain is removed for medical reasons, the right brain is still plastic enough to deputise for most of the duties of the left.

Creative dialectic
The phenomenon of reversed polarity can even affect the body. In 'mirror twins', one twin grows a heart on the opposite side, possibly the result of late cell division in the egg. If nature leaves the separation process too late, the result is Siamese twins, so differentiation is essential.

This does not mean that hemisphericality is a clash of rationality and irrationality, or logic and emotion. It is a dance of partners, of give and take, strength and suppleness, which works only if left brain measures the steps while right brain feels the rhythm.

The phenomena of déjà vu and reincarnation in humans might be explained by micro-second misfiring or delay between hemispheres, leading us to think we are seeing something for the first time, when in fact we are rehearsing previous experience.

Migraine derives from *hemikrania*, meaning 'half brain', experienced as zigzag lines in our left or right visual field following a temporary reduction in oxygen supply to the visual cortex. Some put the phenomenon down to an imbalance of serotonin which constricts blood vessels in one hemisphere, resulting in a thumping

206

headache when blood pressure returns to normal twenty minutes later.

Migraines can be totally debilitating, women three times more likely to suffer from them than men, with no obvious adaptive purpose, though the occasional headache might be a price worth paying for the benefits of doubling-up. Two is better than one, giving us a double perspective and back-up option, as we discover with eyes, lungs, arms, legs, ovaries, testicles and kidneys.

Duality can however generate conflict as well as complementarity. Polarity may even go back to our cosmic beginnings: the electric force works by a tension of plus and minus charges that always seek to be in balance.

Descartes reasoned that the pineal gland in the centre of the brain is the seat of self because it is the only organ not duplicated in each hemisphere. He was wrong about both the role of the pineal gland and the symmetry of the brain. In nature, asymmetry can be a source of strength, adaptation and diversification. Through a creative dialectic, opposites can generate an energy that drives nature forward. The faces judged by both sexes to be most beautiful are usually slightly asymmetric, suggesting that the eye is enticed by difference, not sameness.

The left lung is smaller to make way for the heart, and one testicle is always larger than the other. Quantum physicists tell us that particles possess a positive/negative asymmetry with left and right 'spin', which means that chirality or handedness is an intrinsic bias in nature, often privileging the right hand/left brain. Climbing runner beans spiral to the right, and bees have a bias to the right when scouting for new nest-holes.

Symmetry breaking in nature can be a source of creativity. The DNA double helix has a right bias, but proteins are left-handed. It can also be a matter of life and death. The left-handed molecule of the drug thalidomide can relieve morning sickness during pregnancy, but the right-handed form can cause deformity in the embryo.

If everything were constantly in balance, there would be no incentive for things to change. In human affairs, there would be no yin and yang or saints and sinners, meaning there would be no history, intrigue or progress. Conflict and disruption can be productive, leading to cultural and intellectual breakthrough, but as

207

We Are More Than Our Brains

we shall see, it can also cause dangerous tensions and unnecessary divisions.

How do we integrate east and west?

Duality – polarity - complementarity – thought and feeling – integration

- In our evolution, two halves working together proved to be biologically stronger than one.
- That said, our psychology is shaped by strong rhythms of approach and withdrawal.
- The trick is not to allow the natural divisions in our brain and behaviour to lure us into false opposites in our thinking, or to reinforce cultural prejudices.
- We are at our best when we integrate the spontaneity of our right brain with the caution of our left.
- We benefit physically and mentally by giving ourselves a 'whole brain' workout, not favouring one side or the other.

At loggerheads

If left/right polarity were merely a matter of biological preference or efficiency, we could leave it there, but our cognitive split has cultural consequences too, some of them harmful. We have evolved to think in mutually exclusive either/or terms, such as friend/foe, good/evil, real/imagined. We conduct our politics adversarially, based on a rhetoric of winning and losing, them and us. In discussing brain styles, we talk of thinkers and feelers, convergers and divergers, verbal and visual types. This is not always good news for neighbours who need to discover what they have in common, not what divides them.

We can be left in two minds, sometimes seeing the same thing two ways. 'A fool sees not the same tree as a wise man', remarked William Blake, alerting us to the subjectivity of perception. Functionally however our two halves work in harmony, which is just as well: we have evolved to experience a single consciousness, and cannot cope psychologically with the idea of two persons in one body. Consciousness is a coalition, not a separation of powers.

Some say it's the right side of our brain that makes us human, while the left side assumes executive control. On a good day, we can't tell the difference, and we operate best as whole-brain thinkers. Mystics achieve an even rarer union of hemispheres by experiencing reality as an unbroken continuum, like being absorbed into the misty landscape of a Chinese watercolour painting, becoming one with the Tao or Way. Buddhist monks seek nirvana by learning to calm their frontal lobe activity so effectively that they obliterate the boundaries within their brain, making their mind one with the world.

Few of us achieve such a fusion of realities, though we have no difficulty moving between them quite freely. When we hear our national anthem, we respond both bicamerally and holistically. Our logical left brain recognises the tune, while our emotional right is stirred by patriotic pride in our national symbol. When we watch a movie, we have to be simultaneously analytical and intuitive, 'reading' the screen as both literal signs and implied significances.

Some stutterers, who are statistically more likely to be left-handed, have been cured of their hesitant talking (left) by getting them to sing the words (right). We can love music (right) without understanding its theory (left). Some critics are brilliant analysts (left) but never feel the joy of what they are criticising (right). We can have a touchy-feely moment (right) without being able to explain it (left). Intuition works a bit like this: we know we are right (right) but can't say why (left).

The left brain is tightly packed, proving excellent at fine reasoning, but the right has longer connecting axons so it can appreciate the big picture. If the left sometimes can't see the wood for the trees (too focused on specifics), the right often can't see the trees for the wood (can't itemise the big picture).

We operate at our best when we harmonise both halves of our brain, seeing the foreground clearly against a framing landscape, like a wild animal foraging. We are more likely to get this feeling when we are relaxed, and our more subtle right hemisphere is given equal rein. After a heavy day, we allow the left side some time to recuperate by indulging in some right-side free time.

We Are More Than Our Brains

Two are stronger than one

We can only guess why our brains evolved hemispherically. Two super-connected brains probably offered a more diverse response to reality, and made more space for other functions in each half, adding to the flexibility of an increasingly complex organ. Like having a back-up generator, the brain provides itself with extra protection against the unexpected, or when demand increases.

One theory is that division of labour is more efficient. Birds use one eye to pinpoint food close up and the other to keep a weather eye out for predators (left/specific v right/general view). Pigeons see a separate world through each eye. We humans use our extra brain power and forward-facing eyes to integrate both fields of view, helping us to cope with a more complex social environment. We can have one eye on small detail (left brain: what is she asking me?), the other on the big picture (right brain: can I trust her?).

Our communication with each other is a left/right balancing act. In a conversation we can integrate the words we hear (left) with the facial expression we see (right), combining a cognitive with an emotional reading. We often forget a name (left) but always remember a face (right). When we listen to poetry, the musicality of the language (right) matters as much as its meaning (left).

Living together demands a big brain, but attention-dominating left language tends to overwhelm the quieter right, generating adversarial rather than complementary thinking. If the asymmetry of brain function is the evolutionary price we pay for our cleverness, it is a high one if it limits us to conflicting relations (I'm right/you're wrong) or exclusive ideologies (matter is real/mind is illusory).

Avoiding false opposites

Some attribute spit-brain thinking to a fundamental divide between mindsets, such as Jerusalem versus Athens, religion versus science, feeling versus reason, tenderness versus toughness. The first in each pair is given to belonging, narrative, intuition and imagination, the second looks for separateness, explanation, analysis and practicality.

In the world of culture and ideas, these opposites do not need to be seen as irreconcilable. In the ancient world, Jewish and Greek cultures were not ideological adversaries, but complex mixes of the spiritual and the rational, out of which has grown our modern world, where science thrives within a strong moral tradition. Two millennia later, we do not expect children to experience an existential or intellectual crisis as they proceed from a humanities class to a science lesson at school. Instead we teach them how to look both ways, integrating complementary ways of knowing and experiencing the world.

This is not to say that certain dualisms such as on/off are not a matter of life and death for an electrician. Our problems begin when we allow simple black/white differentiation to spill over into our race relations. While we rely on the left brain's ability to stand back and analyse the fixity of parts, we also need the right's capacity to make us feel we belong to the flow of the whole.

Left and right brain can help each other to recover from a stroke, but we are more likely to survive a left-brain stroke than a right-brain one. Beyond the neuroplasticity of childhood, if the right is damaged, some of its activity can be taken up by the left, but not vice-versa. This is because the left's breezy activities tend to parasitise the right's hard-won experience, which explains why damage to the left hemisphere results in depression (we can't carry out our normal routines) while damage to the right leads to despair (we don't see the point of them).

Approach or withdraw
The brain hemispheres also play different roles in how we deal with emotion and trauma. Remembering that the right brain controls the left side of the body, and vice versa, experiments show that the right brain is better at spotting the difference between a genuine smile, in which the left of the mouth curls up slightly more, and a fake one, where the right does.

Frightening images and memories are lodged in the right, eclipsing the language facility of the left. This shut-down of the rational brain prevents trauma victims from talking through their bad experiences, and explains why soldiers returning from war don't want to talk about what they went through. They are prisoners of their right brain, their emotional clock stopped at the point of trauma. If they are unable to undertake a 'talking cure', their emotions remain locked in. On the other hand, by not talking about them, they let sleeping dogs lie.

We Are More Than Our Brains

The left tends to be bright and optimistic, with an urge to approach, just wanting to get on with things, dealing with explicit memory, or the stuff of everyday. The broody and pessimistic right is more cautious, prone to street-wise misgivings, housing implicit memories of secret fears, deep hurts and simmering grudges. A healthy memory weaves the past and present together as a future-directed narrative, but a traumatised right brain is randomly assailed by buried and unbidden fragments of fear and horror, making it hard to appreciate ordinary feelings or enjoy physical intimacy.

The trick is to hold the positive left and negative right in balance. Some left-brainers are over-confident about their abilities, deaf to the warnings of their right brain, while some right brainers, blind to the opportunities on offer, navel-gaze too much and never take chances. Somewhere between being too cocky and too moody there is a happy medium, and we need both impulses if we are to learn.

We process the unfamiliar in our right brain before it is passed to the left as safe, but if we don't take risks, we're stuck with what we have. It's one thing to have an angry gut reaction (right), another to work out what is causing it (left) so that we can offer a more measured response next time.

Anxiety has a useful role if it prevents us from getting hurt or making fools of ourselves. Intelligence and social success are defined by our ability to make the 'right' call when facing a new challenge. The left is keen to process reality, the right is on guard for what may harm us.

Our bad feelings about ourselves tend therefore to reside in the right, its image-obsessed nature struggling to erase painful memories, but cognitive bias modification is a therapy that uses the left brain's reasoning potential to rethink attitudes and rationalise the pain: it's not the spider we fear, but our prejudices about spiders. We can push back the negative thoughts emanating from the right brain, and capitalise on the left's cheerfulness, but only by conscious effort and persistent practice.

Staying in touch with our feelings
We should be wary of generalisations, but left-brainers tend to be happy-go-lucky, pulling us through with their positivity. They also stick with the familiar, working well with what they know, their

insight and creativity excelling in technical innovation, systems analysis and the discovery of physical realities.

Right-brainers fare better with the expression of feeling, 'seeing' with their emotions, making art out of what was not there before. Left-siders are creative and imaginative too, but their breakthroughs tend to come, not after hours of disciplined thought, but when they 'sleep on' their ideas, allowing their wacky right brain to work its magic.

We often excel at drawing as children because we see the world holistically, our left and right brain ways of seeing not fully differentiated. As soon as we start to look only with our left brains, as detached and analytical agents, we lose our confidence, and cease to feel the unity of the whole. In adulthood we need to stay in touch with our feelings, otherwise we lose our right-sided gateways to intuition and spiritual renewal. The great artists of the world keep us in touch with our hidden side, reminding us who we are deep down.

If we could disconnect our moody right side, we'd be permanently elated, but it would cost us our personality, and happiness would become meaningless, as all emotions would lose their edge. We would also place ourselves in mortal danger, because our emotions inform our reason and keep us sane. Our intellect is capable of building a super-intelligence, but only our emotions can caution us about where our ingenuity might lead us.

Left and right are not rival but complementary modes of experience. If our divided brain forces a split between thought and feeling, or our culture of 'hard' sciences and 'soft' arts forces mutually exclusive left brain/right brain ideologies onto us, it is all the more important to nurture an even-handed response to both to balance the bias.

Life exposes and intensifies our lop-sidedness soon enough. We fare better at some things than others, because few of us are born ambidextrous or multi-talented. We are not like the gifted artist Landseer who could simultaneously draw a different animal with each hand, or a talented footballer who can effortlessly score a goal off either foot.

Despite the fact that few of us are both-handed, and most of us are right-handed, we do not need to turn the dominance of the right into a cultural prejudice or ideological divide: life has enough self-inflicted dualities without adding to the list. Rather than handicap ourselves by pensioning off half our brain, we need to make it work as hard as the other half, giving our brain a workout by occasionally doing a job with our 'wrong' hand, if only to remind us how skilled our 'good' hand is.

Psychometric testing and criminal profiling do us no favours by itemising our handedness, IQ, psychological strengths and character weaknesses. We are not so easily 'typed', pigeon-holed or fitted for a role in the work place or society. A rounded character, balanced mind and even-handedness are needed to succeed in the business of living.

How do two brains become one mind?

Orthodoxy – heterodoxy – creativity - handedness – mosaic mind

- Since ancient times, society has been institutionally prejudiced against left-handers.
- This bias has also extended to unorthodox or 'left field' thinking, quashing innovation and eccentricity.
- Right-brained dreaming has however led to significant breakthroughs in all disciplines and walks of life.
- Social and political crises have arisen when left or right brained thinking is pushed to extremes, preventing one from tempering the excesses of the other.
- The ancient Greeks saw the life of the mind as a harmony of logos and mythos, or reason and imagination. Our psychic health suffers when these are out of kilter.
- In our evolution, genes for right-handedness have been in the ascendant.
- Left-handed tools have been found in the archaeological record, but by dint of numbers, right-handed tools became the standard.
- This does not mean that written scripts have to read from the right, or we have to drive on the left. These are cultural norms, not biological diktats.
- In recent times, the return to a predominantly right-brained visual culture has tempered the 'wordiness' of left-brained thinking and restored the mosaic of experience.
- What unites all our ways of seeing and knowing is their reliance on symbols, which are the greatest achievement of human thought.

We Are More Than Our Brains

Right thinking

Handedness is more than physiology. It has taken on moral overtones, loaded with cultural prejudice, erecting barriers that divide us. The dominance of right-handedness is equated with being morally right, a sign of trustworthiness and authority. We do the right thing, we know our rights, we are pleased to sit on the right hand of God. Usually the right is proffered for a handshake, so a left handshake feels odd. These are all cultural biases against the ten per cent of the population who are left-handed.

Even retailers cash in on our right-handed proclivity. Psychologists tell us that in experiments where volunteers are asked to choose from four identical items laid out on a table, they predominantly choose the one on the right. If left-handers fall in with this pattern, they are probably responding to a subconscious desire to conform.

In a charming display of biological ignorance as well as male chauvinism, the Ancient Greeks believed that by tying a ligature around the smaller left testicle, weak female seed would be held back, and stronger male seed could be guaranteed to flow from the larger right testicle.

Cultural practices also favour the right. In Arab countries, food should always be offered with the right hand, because the left is reserved for personal cleanliness. At sea, it is a calculated insult by the captain to force anyone to disembark from the port or left side of his ship. As a contrast, Egyptian statues always stand with their left foot forward, on the side of the heart, because that is a symbol of life.

Despite the occasional exception, the anti-left conspiracy runs deep, through our language and into our thinking. Dexterity (Latin for right) is manual skill, but the sinister (left) is deformed. Orthodoxy (Greek for straight) is right thinking and approved teaching, but heterodoxy is twisted 'other' opinion, to be rooted out as heresy.

Such false opposites of language have affected our attitudes to sexual inclination for centuries. We still call those who conform to the norm 'straight', imposing the shadow terms 'pervert', 'bent' and 'queer' on those who don't fall within the statistical norm.

In the western tradition, right-handedness and left-brain thinking are seen as the legacy of the sun-god Apollo: useful, reliable,

217

logical and analytical, emblematic of rational mind. Left-handedness and right-brain thinking are by contrast the mark of the night-owl Dionysus: dark, dangerous, unpredictable and irrational, an affair of the blood which encourages emotions to run high.

LEFT Count the trees	RIGHT See the forest
Analysis of specifics	Integration of the whole
Perception	Sensation
Sequences	Correlations
Scrutiny	Hypothesis
Knowledge	Beliefs
Quantity	Quality
Calculation	Judgment
Words	Images
Grammar	Meaning
Clarity	Ambiguity
Logic	Emotion
Facts	Intuitions
Familiarity	Newness
Breezy	Brooding
Curious	Cautious
Deliberate	Automatic
Waking	Dreaming
Disassembling	Reassembling
Sequential	Networked
Critical	Speculative

The left hemisphere evolved to help us count the trees, but the right enables us to see the forest. We may have a cognitive style that sports one side of the brain more strongly than the other, and in some brains these tendencies are reversed, but we need both to get the full picture.

We Are More Than Our Brains

In French, 'droit' means not only right but also law (royalty's motto is 'Dieu et mon droit'), and maladroit means clumsy. 'Gauche' means left, and is used in English to denote awkwardness. Greek joins the anti-left conspiracy too: *kakos*, meaning bad, gives us cack-handed left-handers.

The English word left comes from Anglo-Saxon 'lyft' meaning worthless, the work of the weak hand. 'Left field' ideas from the left hand are dismissed as unconventional to the point of irrelevance. Numerical evenness and oddness also find their way into our moral thinking: an even judgment is right, but an odd decision is faulty.

Left field thinking

So it seems our brain laterality carries with it a lot of design faults that manifest themselves as ideological baggage, ensnared with dualities of saint/sinner, friend/foe, clean/polluted, logic/love and reason/irrationality, most of which are not mutually exclusive when we take time to unpick them. In an ambiguous world, these false opposites are moral handicaps and intellectual pitfalls. None of us is a pure strain of either/or, because we are all complex blends of both/and.

As with all dualisms, there are many counter-factuals. Excessive left thinking can be arrogant, single-track, isolating and fragmenting, and left brain can be so objectifying as to be judgmental, impersonal, chauvinist and reductive. Cocky stockbrokers trade (and lose) millions based on 'logical' left-brain calculation, deaf to the quieter caution and hunch of right brain. Money is an expendable loss/gain commodity, but for our more complex social relations we need the diverse, plural, tolerant, inclusive, receptive and humanising influence of the right brain.

The psychologist Julian Jaynes believes it was the intuitive voice of the right brain that broke into the fixed routines of the left brain about five thousand years ago to give us our modern sense of self, though we could just as easily argue that the left broke into the right.

Feminists claim that it was right-brain female subtlety and sensitivity that started to break down male control and aggression. We don't hear of many female investment bankers bringing the

219

system crashing down, and we could certainly do with more caring and ethical business practices in the boardroom.

Right-brained dreamers are accused of never getting anything done, but in a world ruled by orthodoxy and right-handedness, we need the alternative perspective, radical approach and creativity of heterodox right-brained left-handers, male and female, such as Joan of Arc, Leonardo, Newton, Beethoven, Marie Curie and Charlie Chaplin, who all came up with 'left field' ideas.

We must beware however of over-applying an ideology of handedness. It is often claimed that Einstein, Picasso and Bob Dylan were left-handed, but there is no evidence for these claims. Most of the 'alternative' right-brained (but not necessarily left-handed) Flower-Power Children of the 1960's went on to become more conventional and conservative than the (mostly male) left-brained 'squares' they once rebelled against.

Creative bursts

Some cognitive neuroscientists credit our right brain with the secret of our creativity. Its urge to communicate in images leads to 'eureka' moments for both artists and scientists when, after a period of tightly-focused 'top-down' analysis of a topic, which may last years, they allow more loosely connected and unusual thoughts from their 'intuitive' unconscious to break into their 'analytic' conscious mind, as in sleep, reverie or relaxation. Researchers have even tracked down the 'aha' moment of insight to a burst of gamma wave activity just above the right ear.

Wherever inspiration occurs in the brain, mysticism, philosophy, literature, art and science are full of stories of right-brain dream visions. The poet Dante claimed the vision for his 'Divine Comedy' came to him in a dream, the paintings of Hieronymus Bosch seem to lure us into a waking dreamscape, Descartes' philosophical *cogito* 'I think therefore I am' came to him in a nocturnal reverie, and the chemist Kekulé's realisation of how atoms bond into molecules popped into his mind as he sat musing on a bus. He also remarked however that he had to apply his left brain when he got home to turn his daydream into a workable theory.

Many modern physicists claim deeper understanding of the cosmos through a kind of mystical insight. Scientists, usually associated with dominant left-brain analytical thinking, tell us that some of their greatest work emanates from right-brain moments of doubt, intuition and artist-like inspiration, as a kind of daytime dreaming. 'The great hypotheses of science are the gifts carried in the left hand', wrote the cognitive scientist Jerome Bruner.

We are wrong to dismiss right-brained creative types or artists as scatty geniuses: most plan and order their work as scrupulously as any scientist conducting an experiment. The 'creative scientist' and 'methodical artist' are not oxymoronic contradictions but fruitful partnerships, both requiring imagination, which arises from whole-brain connectivity. Left and right are equally valuable thinking styles, and true intelligence resides in knowing when each is called for or how to integrate them. Our great achievements of language, art, science and religion are collaborations of left and right, not monopolies of one domain.

History offers us many lessons of what happens when we divorce thinking from feeling. Too much left brain can leave us with dull efficiency and lower levels of contentment, sidelining myth, art and religion as anodyne leisure activities. Too much right brain can result in extravagant social experiments, reduced liberties and false gods. When pushed to extremes, both ideologies are haunted by the spectres of revolutions, tyrannies, mutual destruction, gulags and death camps.

Science and the arts can together reward us with the intelligence of feeling and the joy of discovery. Immanuel Kant sought to unify his knowledge and understanding by retaining both his wonder at the starry skies above (left-brained public science) and the moral law within him (right-brained personal responsibility).

Celebrating difference
Genes for right-handedness won out over left in our evolution, wiring our brains in a certain way. Two left-handed parents are still statistically likely to produce a right-handed child. Most mothers hold their babies on the left, next to their heartbeat, in closer contact with their right-brain emotional processing and freeing their right hand to stroke the child.

There are however no biological grounds for condemning the left as inferior, only cultural ones. Right-handers on average live a few years longer, and experience fewer learning problems, but the

221

danger of generalisations is that they mask a host of exceptions. In any case, archaeologists have found left-handed Stone Age axes, so difference and unorthodoxy are written into our DNA.

One in ten of us is a 'southpaw', so there must be a reason for left-handedness surviving in the gene pool. One theory is that left-handed warriors had the edge over right-handed rivals on the battlefield. Holding their shield on the right, and their sword in their left hand, they literally came in from a different angle, wrong-footing their opponent.

This might explain why left-handers are well represented in sports like boxing, tennis, golf and cricket, winning their share of prizes. It also accounts for the presence of left-handed scissors, guitars, saws, pianos and golf clubs. Partners who are opposite-handed have to come to a 'middle ground' compromise: always leave the kettle with the handle in a central position.

'Drive on the left' is not a natural inheritance but a cultural convention. We can drive on the right if we wish, so long as we all remember to do so. There may however be a biological explanation that originally favoured keeping to the left. Right-handed horse riders tend to mount and dismount the horse from the left, and want their stronger sword arm free to face strangers approaching from the right. In jousting, the knight wants his lance under his stronger right arm as he charges towards his adversary. Nevertheless, Napoleon's decision to drive on the right was a left-field political one, to assert a new revolutionary order, and no horses were consulted in the process.

Left-brained writing and logic dominate our culture. The images of the right can't 'talk', so are pushed into second place in our schools, regarded as slightly suspect. Our culture looks down on left-hand/right-brain touchy-feeliness. College art courses are likely to be axed before 'useful' science and technology, but we need both if we are to live the good life. Socrates and Jesus relied on right-brained face-to-face oral teaching, but our school examinations are dominated by the left-brained methods and writings of their apostles, Plato and St Paul.

Cultural favouritism
Western civilisation inherited from the Ancient Greeks a strong bias towards left-oriented logic and language, and a mistrust of right-

We Are More Than Our Brains

brained emotionality and images. Some speculate that the adoption of the alphabet encouraged the linear thought of left-brained reason, ushering in the scientific revolution, but without order and syntax, right-brained poetry could not be written either.

Handedness is not fixed in the womb until about fifteen weeks, and parents cannot tell for many months after birth which hand their child favours, usually accepting it without demur. Cultural favouritism of right-handedness has however been responsible for centuries of prejudice against left-handers. Cruel punishment was inflicted on unorthodox children in times past, their left hand tied behind their backs until they were 'cured' of their 'sinister' habit.

And yet there is nothing intrinsically 'right' about right-handedness. As a species with two hands, we have to distinguish between left and right, but that doesn't equate to right and wrong. Some children struggle at first to distinguish *b* and *d* in their writing, but they learn the 'right' way round for the letters eventually by dint of repeated encounter.

Writing from left to right is favourable to right-handers but awkward for lefties who complain that their writing hand smudges the ink or obscures the words they have just written. Western civilisation has not always started writing a page in the top left corner, proving that doing so is a social agreement, not a rule of the universe. Why then did left-to-right win out over rival systems?

Semitic languages such as Hebrew and Arabic are read from right to left, meaning that when we read the Torah or the Koran, the back cover is in fact the front cover. The Semitic alphabet also contains no vowels, obliging the reader to engage more actively with the context and meaning of each word.

The Greeks borrowed the Semitic alphabet, but made several important modifications. They added vowels and wrote 'as the ox ploughs', making the eye track back and forth across the text. Eventually they opted for left to right script, and inserted spaces between words to assist reading.

These left-right cultural differences between alphabets and writing conventions lead some to suggest that Semitic scripts give a more holistic understanding of what is being read, because they work the right brain harder. This makes them ideal for story-telling of a mythical kind. The more analytical Greek approach by contrast activates the left brain, giving a kick start to philosophy, science and the writing of history based on secular evidence, not sacred narrative.

No-one can agree whether the script we are taught restructures our brain or consciousness in anything but a superficial way. How we learn to read and write are not claims to unique ways of experiencing the world: the underlying thought processes are the same. If we add the fact that Chinese script is written not with letters but 'painted' with ideograms that run down the page, the issue becomes even more complicated.

Yod Tet Chet Zayin Vav He Dalet Gimel Bet Alef
(Y) (T) (Ch) (Z) (V) (H) (D) (G) (B/V) (silent)

Ayin Samech Nun Nun Mem Mem Lamed Khaf Kaf
(silent) (S) (N) (N) (M) (M) (L) (Kh) (K/Kh)

Tav Shin Resh Qof Tsadeh Tsadeh Feh Peh
(T) (Sh/S) (R) (Q) (Ts) (Ts) (F) (P/F)

The Hebrew alphabet
Note that alef, or A, is top right. Choosing to read from right to left or vice versa is determined by culture, not by how the brain works or is divided.

Perhaps the differences these conventions make in our brain are as trivial as deciding whether men should button their clothes from the right and women from the left. After all, clocks do not *have* to go clockwise, though the habit may arise from time-tellers in the northern hemisphere, where clocks first appeared, looking south and tracking the sun to the right from its rise in the east to its setting in the west. Clocks could run anti-clockwise if we wanted them to. Similarly, deaf people using sign language can choose to 'lead' with either hand. All that matters is that they are consistent.

On the other hand, it is a matter of necessity that water runs down the plughole in opposite directions in each hemisphere, and straight down at the equator, because it is pulled by the physical force of gravity, against which there is no argument. There is no such necessity to place the knife on the right when we are laying the table, especially if our guest is left-handed. Then again, having cut up their food, Americans transfer their fork to their right hand, so they eat 'on the other hand' regardless.

Mosaic mind
The media theorist Marshal McLuhan saw in modern culture a 'mosaic' model of mind which resolves left/right duality, as right-brained non-linear pictures regain the hold that they once had over the mind in the preliterate Middle Ages, challenging the left-brained linear supremacy of language.

If he could see the proliferation of images, graphics, icons, logos and hypertext in today's digital culture, he might think he was right, but whether we are cleverer, better informed, more bewildered, less illuminated or more finely educated as a result of the resurgence of the image and the retreat from the printed word is a different question altogether.

The appeal of the image has always been strong in popular culture. Mediaeval priests knew that bright murals on church walls fast-feed the gospel to the image-sensitive right brain more effectively than dry sermons, and modern advertisers know we are greater suckers for evocative pictures and stories than we are for tedious fact-based sales-talk. Film is a popular art form because its images can be taken in whole by the right hemisphere. Newspapers sell better if they place images and graphical information to the left of a page, because such information is fed directly to the shape-sensitive recognition modules in the right brain.

Reading is more cognitively demanding than watching because the left-to-right flow of our writing means letters are 'seen' in the right first, then have to be decoded in the dominant word-processing left.

Neural logjam
A unified brain is a successful brain: we perform better when we co-ordinate our actions and think ahead. In the field of technology, a whole brain is better than a half if we are to consider the full impact of innovation on human lives.

Many of us find we work better (left) with music in the background (right) to soothe and stimulate creative flow, allowing our ideas to seep through porous membranes. We can perform two tasks at once so long as we don't choose conflicting pathways or

225

call on each side twice in one operation. If we're phoning while we're driving, we're heading for a crash: the left-brain demand to talk into the machine reduces right-brain awareness of the road ahead by a third.

When we try simultaneously to circle our right arm clockwise and our right leg anti-clockwise, we become quickly aware of our brain's limitations in handling mixed messages in the left, but the right is better at multi-tasking. Composers sometimes run one tune over the top of another, something the holistic right brain can cope with, but trying to run two conversations at once jams the linear capacity of the left, as we discover when there is a two-year-old in the room. The left needs a single focus while the right revels in variety. We can process visual input from split screens, but only one soundtrack.

What unites both sides of the brain is the power of the symbol: words can express clinical logic (the heart is a pump) or melancholic emotion (my heart aches). But symbols mean nothing except as acts of mind in a shared culture. We are more than our brains. As well as grasping our neural history and mental geography, we need to understand how it is our mind that makes sense of what our brain gives us.

Knowing our own mind

Neuroscience answers the call of an ancient imperative, to know our own minds. The motto 'Know Thyself' was inscribed in the Temple of Apollo at Delphi, and according to the Chinese sages, 'Not to know our own mind is to miss out on a treasure that is ours'.

Previous ages have been locked out of the brain's hidden workings, and mystified by the nature of mind, but courtesy of neuroscience, we can now see ourselves through the prism of new knowledge. It is as if, having lived all our life in one place, neuroscience has allowed us to know ourselves fully for the first time.

We have been given a new language to talk about ourselves: neuroconnectivity, neuroplasticity and neurodiversity. And yet when we reflect, we realise that these are old truths dressed up in new clothes.

Neuroconnectivity   In our brain, as in life itself, everything is connected to everything else, all things working together to create more than the sum of their parts. We understand nothing unless we join up all our ways of thinking and endeavour to see things whole.

Neuroplasticity   Aristotle was the first to suggest that we possess *metanoia*, or the power to reshape our brain, remake our mind and have a change of heart. Within certain limits, neuroscience has vindicated him. So long as we stay creative, learning can be a lifelong process, and we can keep the future open.

Neurodiversity   Evolution grants us particular gifts, and life writes customised scripts for us. Diversity is essential for human flourishing, giving our personality its individuality and society its vitality. Our differences are to be expected, accepted and celebrated.

We Are More Than Our Brains

We are our brains, in the sense that we are clever survival machines. This is what a brain is for.

But we are also our minds, in that we make meanings, navigate our complex relations with each other and imagine a better future. This is what a mind is for.

Further Reading

General introductions to the mind-brain

Blakemore, Colin – *Mechanics of the Mind* 1977   How we get mind from the mechanics of brain.

Cobb, Matthew – *The Idea of the Brain* 2020   A sociocultural history of our understanding of the brain and how it works.

Eccles, C and Popper, K – *The Self and Its Brain* 1984   A scientist and a philosopher compare their views about the mind-brain.

Edelman, Gerald – *Bright Air, Brilliant Fire* 1991   An enthusiastic manifesto for neuroscience and its ability to explain the matter of mind.

Gazzaniga, Michael – *Human* 2005   The science behind what makes the brain unique.

Flanagan, Owen – *The Science of Mind* 1984   Is a science of mind possible?

Frith, Chris – *Making Up The Mind* 2007 How our brain creates our mental world, which is not always what it seems.

Greenfield, Susan – *The Human Brain* 1997 A guided tour.

Horgan, John – *The Undiscovered Mind* 1999   How the brain defies explanation.

Le Doux, Joseph – *The Deep History of Ourselves* 2019   An evolutionary perspective on how our mind-brain has taken shape.

Lieberman, Jeffrey – *Shrinks* 2015 A psychiatrist's perspective on how psychiatry has gradually taken on the lessons of neuroscience.

McGilchrist, Iain – *The Master and his Emissary* 2010 How the structure of the brain has shaped modern thought.

New Scientist – *Your Conscious Mind* 2017   Unravelling the greatest mystery of the human brain.

Ornstein, Robert – *Multimind* 1986 How our mind is not single but modular.

O'Shea, Michael – *The Brain* 2005 A very short introduction.

Pinker, Steven – *How the Mind Works* 1997 Mind from an evolutionary psychology perspective.

Robertson, Ian – *Mind Sculpture* 1999   Exploring the untapped potential of the brain.

We Are More Than Our Brains

Rattray Taylor, Gordon – *The Natural History of the Mind* 1979 Dated but still topical.

Sapolsky, Robert – *Behave* 2017 A magisterial study of the biology that makes us who we are, warts and all.

Smith, Anthony – *The Mind* 1984 Companion to his earlier 'The Body'.

Sternberg, Robert – *In Search of the Mind* 1998 A wide review of the psychology of the mind-brain.

Winston, Robert – *The Human Mind* 2003 How to understand and make the most of what we have between our ears.

The perspective of neuroscience

Begley, Sharon – *The Plastic Mind* 2009 How our brain can adapt, renew and transform itself.

Bentall, Richard – *Madness Explained* 2003 A study of psychopathy and its relation to the healthy mind-brain.

Bergland, Richard – *The Fabric of Mind* 1986 From brain hormones to mind.

Blakemore, Sarah-Jayne – *Inventing Ourselves* 2018 The secret life of the teenage brain.

Bullmore, Edward – *The Inflamed Mind* 2018 The links between an overactive immune system, an inflamed brain and depression.

Burnett, Dean – *The Idiot Brain* 2016 What our head is really up to.

Buzacki, Gyorgi – *The Brain from Inside Out* 2019 How the brain is not a passive processor of inputs but an active maker of reality.

Carter, Rita – *Mapping the Mind* 1998 From brain scans to culture and behaviour.

Challoner, Jack – *The Brain* 2000 Ideas to accompany the Channel Four Equinox investigations into neuroscience and the brain.

Changeux, J P – *Neuronal Man* 1985 An eminent neurobiologist goes in search of the biology of mind.

Chater, Nick – *The Mind is Flat* 2018 A strong argument that depth of mind is an illusion, and that our brain makes things up as it goes along.

Churchland, Patricia – *Touching a Nerve* 2013 The ethical and practical implications of brain science.

We Are More Than Our Brains

Churchland, Paul M – *The Engine of Reason* 1995 The social dimension of brain science.

Critchlow, Hannah – *The Science of Fate* 2019 A study of the ways in which our future is already determined by our biology and brain activity.

Damasio, Antonio – *Descartes' Error* 1994 The role of emotion in reasoning.

Domingos, Pedro – *The Master Algorithm* 2015 How attempts to reverse-engineer the brain to create the ultimate AI learning machine provide many insights into how our grey matter pulls it off so effortlessly.

Doidge, Norman – *The Brain that Changes Itself* 2007 Case studies of the power of the brain to self-renew.

Dweck, Carol – *Mindset* 2006 Our power to change the way we think.

Eagleman, David – *The Brain: The Story of You* 2015 An account of how our brain shapes our identity and awareness.

Evans, Dylan – *Emotion* 2001 The importance of emotion in shaping our cognitive landscape.

Frank, Lone – *Mindfield* 2009 How brain science is changing the world.

Frith, Chris – *Making Up The Mind* 2005 How the brain creates our mental world.

Goleman, Daniel – *Emotional Intelligence* 1996 Why EQ (emotional intelligence) can matter as much as IQ.

Hacking, Ian – *Rewriting the Soul* – 1998 Does multiple personality disorder have a neurological base, or is it an invention of psychiatry?

Horgan, John – *The Undiscovered Mind* 1999 Testing the claims of neuroscience.

Humphrey, Nicholas – *A History of the Mind* 1992 An account of the evolution of consciousness.

Jaynes, Julian – *The Origin of Consciousness* 2000 A theory that consciousness arose from the 'breakdown of the bicameral mind'.

Johnson, Steven – *Mind Wide Open* 2004 Why we are what we think.

Lehrer, Jonah – *Proust was a Neuroscientist* 2007 What the arts have long taught us about the mind.

We Are More Than Our Brains

Linden, David – *The Accidental Mind* 2008   How love, memory, dreams and God are 'accidents' of evolution.

Luria, A R – *The Man with a Shattered World* 1974   A historic case history of a victim of a serious brain wound.

Marcus, Gary – *Kluge* 2008 How our mind is not 'designed' but the product of haphazard evolution.

McManus, Chris – *Right Hand, Left Hand* 2003 Reflections of how brain laterality has played out in human culture.

Mlodinow, Leonard – *Subliminal* 2013   How our unconscious mind rules our behaviour.

Newman, Robert – *Neuropolis* 2017   The inadequacy of neuromania as a theory of everything.

Passingham, Richard – *Cognitive Neuroscience* 2016   A pithy account of what neuroscience can and cannot tell us.

Ramachandran, V S – *The Tell Tale Brain* 2011   Lessons from when the brain goes wrong.

Ropper, Allan – *Reaching Down the Rabbit Hole* 2015   A clinical neurologist takes us on extraordinary journeys into the human brain.

Rose, Steven – *The 21st Century Brain* 2006 The virtues and vices of neuroscience.

Sachs, Oliver – *The Man who Mistook his Wife for a Hat* 1985 Case histories of neurological disorders.

Satel, Sally – *Brainwashed* 2013   A warning against over-applying the findings of neuroscience.

Schwartz, J and Begley, S – *The Mind and the Brain* 2002   The importance of neuroplasticity in therapy.

Scruton, Roger – *On Human Nature* 2017   A philosopher explains why brain is necessary but not sufficient to account for the personal qualities of mind.

Seung, Sebastian – *Connectome* 2012   How the brain's wiring makes us who we are.

Sigman, Mariano – *The Secret Life of the Mind* 2017   The hidden forces that drive our thinking.

Strauch, Barbara – *The Secret Life of the Grown-up Brain* 2010   How the brain gets wiser as it gets older.

Swaab, Dick – *We Are Our Brains* 2014   An account of the brain's role in giving rise to the mind.

Tallis, Raymond – *Aping Mankind* 2015   A philosophical salvo against reductionist materialism in brain science.

We Are More Than Our Brains

Taylor, David A – *Mind* 1983   A scientist's view of how the mind works.

Taylor, Kathleen – *The Brain Supremacy* 2012   Exploring ethical dilemmas at the frontiers of neuroscience.

Thomson, Helen – *Unthinkable* 2018 The effect on the mind when the brain misfires.

Varela, Francisco et al – *The Embodied Mind* 1991   A plea to include the subjective mind in our objective study of the brain.

Warren, Jeff – *Head Trip* 2007   A 'fantastic romp' through a day in the life of the brain.

Zimmer, Carl – *Soul Made Flesh* 2004   An account of the challenges facing the first neuroscientist Thomas Willis four hundred years ago.

Acknowledgement
The frame from 'The Numskulls' is taken from The Beezer comic, a great favourite of mine back in the day. It appeared in the 1960's, drawn by Malcolm Judge and published by D C Thomson.

Index

We Are More Than Our Brains

We Are More Than Our Brains